Stefan Wolf

Silicon-organic hybrid (SOH) electro-optic modulators for high-speed and power-efficient communications

Karlsruhe Series in Photonics & Communications, Vol. 23
Edited by Profs. C. Koos, W. Freude and S. Randel

Karlsruhe Institute of Technology (KIT)
Institute of Photonics and Quantum Electronics (IPQ)
Germany

Silicon-organic hybrid (SOH) electro-optic modulators for high-speed and power-efficient communications

by
Stefan Wolf

KIT Scientific Publishing

Karlsruher Institut für Technologie
Institut für Photonik und Quantenelektronik

Silicon-organic hybrid (SOH) electro-optic modulators
for high-speed and power-efficient communications

Zur Erlangung des akademischen Grades eines Doktor-Ingenieurs
von der KIT-Fakultät für Elektrotechnik und Informationstechnik des
Karlsruher Instituts für Technologie (KIT) genehmigte Dissertation

von Dipl.-Ing. Stefan Wolf geboren in Aalen, Deutschland

Tag der mündlichen Prüfung: 18. Dezember 2017
Hauptreferent: Prof. Dr.-Ing. Christian Koos
Korreferenten: Prof. Dr.-Ing. Dr. h.c. Wolfgang Freude
Prof. Dr.-Ing. Thomas Zwick

Impressum

KIT Scientific Publishing

Karlsruher Institut für Technologie (KIT)
KIT Scientific Publishing
Straße am Forum 2
D-76131 Karlsruhe

KIT Scientific Publishing is a registered trademark
of Karlsruhe Institute of Technology.
Reprint using the book cover is not allowed.

www.ksp.kit.edu

Print on Demand 2018 – Gedruckt auf FSC-zertifiziertem Papier

ISSN 1865-1100
ISBN 978-3-7315-0810-6
DOI 10.5445/KSP/1000084027

Table of Contents

Kurzfassung

Kommunikation ist ein unverzichtbares gesellschaftliches Element und elementarer Bestandteil zivilisierten Lebens. Die modernen technischen Möglichkeiten haben die Art der Kommunikation massiv beeinflusst. Die rasante Entwicklung der Kommunikationsmöglichkeiten hat vielfältigen Einfluss auf fast alle Gesellschaftsbereiche. Die permanente Verfügbarkeit dieser Kommunikationsmöglichkeiten verdanken wir einem globalen optischen Kommunikationsnetzwerk. Dabei sind die Anwendungsfelder in der optischen Kommunikationstechnik und die damit verbundenen Anforderungen an die Technologie vielseitig. Durch den immer weiter ansteigenden Bedarf an Bandbreite ergeben sich ein hohes Maß an Innovationspotenzial und –bedarf: Datenzentren sind ein zentraler Bestandteil des globalen Kommunikationsnetzes. In Datenzentren gelten strenge Anforderungen an die Größe und Energieeffizienz von optischen Kommunikationsmodulen, zu deren Kernkomponenten der elektrooptische Modulator zählt. Diese Anforderungen können von konventionellen Modulatoren auf Basis von Lithiumniobat ($LiNbO_3$) nicht erfüllt werden, da diese zu groß und für die Integration nicht geeignet sind. Aus diesem Grund hat sich in den letzten Jahren die Forschung auf halbleiterbasierte optische Modulatoren konzentriert. Mit der bereits existierenden Fabrikationsinfrastruktur aus der Complementary-metal-oxide-semiconductor-Technologie (CMOS) und der Möglichkeit zur Massenfertigung ist Silizium eines der vielversprechenden Materialsysteme. Gegenwärtige optische Modulatoren aus Silizium beruhen auf dem Prinzip des Plasma-Dispersionseffekts, aber die begrenzte Modulationseffizienz setzt Grenzen in Bezug auf die Energieeffizienz dieser Bauteile.

Die vorliegende Arbeit befasst sich mit einem alternativen Ansatz zur Realisierung elektro-optischer Modulation auf Silizium. Anders als bei konventionellen Siliziummodulatoren macht sich dieser Ansatz die Nichtlinearität zweiter Ordnung in organischen Materialen zunutze. Diese Materialien werden auf den Silizium-Chip zusätzlich aufgebracht. Die Modulationseffizienz sogenannter silizium-organischer Hybrid-Modulatoren (silicon-organic hybrid, SOH) übersteigt die von konventionellen Modulatoren um mehr als eine

Größenordnung. Die Fabrikation von SOH-Modulatoren ist kompatibel mit standardisierten lithografischen Fabrikationsprozessen und erlaubt eine skalierbare kosteneffiziente Herstellung. Die vorliegende Arbeit nutzt SOH-Modulatoren zur Erzeugung von Hochgeschwindigkeitsdatensignalen mit Symbolraten bis zu 100 GBd – Rekordwerte für halbleiterbasierte Modulatoren. Darüber hinaus wird die Anforderung an energieeffiziente Systeme erfüllt: SOH-Modulatoren können ohne elektrische Verstärker direkt von den binären Ausgängen von Field-programmable Gate-arrays (FPGA) mit Spannungspegeln unter 300 mV_{pp} betrieben werden. Höherwertige Modulationsformate werden dabei ohne Digital-Analog-Wandler (digital-to-analog converter, DAC), sondern lediglich unter Verwendung passiver Hochfrequenzkomponenten erzeugt. Mit Treiberspannungen von nur wenigen hundert Millivolt beträgt der Energieverbrauch weniger als 20 fJ/bit.

Kapitel 1 gibt eine Einführung in die optische Kommunikationstechnik und einen Überblick über die aktuellen Herausforderungen in Bezug auf photonische Integration.

Kapitel 2 erörtert die Grundlagen optischer Kommunikation. Es erklärt zunächst optische Signale, optische Modulationsformate und zugehörige Empfangsmechanismen. Der zweite Teil befasst sich mit elektro-optischen Modulatoren und diskutiert die Grundlagen von siliziumbasierten Modulatoren. In diesem Zusammenhang wird das Prinzip von SOH-Modulatoren aufgegriffen und detailliert erörtert.

Kapitel 3 befasst sich mit der Übertragung von On-off-keying- Datensignalen (OOK) mit SOH-Modulatoren. Optische Datensignale mit einer Datenrate von 100 Gbit/s werden mit niedrigsten Treiberspannungen von nur 1.4 V_{pp} und einem elektrischen Energieverbrauch von weniger als 100 fJ/bit erzeugt. Es wird gezeigt und analysiert, dass die hohe Modulationseffizienz und die daraus resultierenden geringeren Anforderungen an die Treiberspannung dazu führen, dass Bandbreitenbegrenzungen des Modulators und der elektronischen Komponenten gemildert werden. Dadurch können optische Signale auch über die charakteristische 6 dB-Bandbreite hinaus erzeugt werden.

In *Kapitel 4* werden Aspekte kohärenter Kommunikation mit Inphasen-Quadratur-SOH-Modulatoren (IQ) behandelt. Es werden Datensignale mit dem Modulationsformat Quadraturphasenumtastung (quadrature phase shift keying, QPSK) und mit Quadraturamplitudenmodulation mit 16 Zuständen (16-state quadrature phase shift keying, 16QAM) bei Symbolraten bis zu 100 GBd erzeugt. Dabei handelt es sich um die erste Demonstration von 100 GBd-16QAM-Signalen mit halbleiterbasierten Modulatoren. Die erreichte Datenrate von 400 Gbit/s wurde bisher mit keinem anderen silizium-basierten Modulatorkonzept erreicht.

Kapitel 5 umfasst Lösungsansätze zur energieeffizienten optischen Kommunikation. Es wird über die Erzeugung von optischen Datensignalen ohne die Verwendungen von elektrischen Verstärkern berichtet. Dabei werden die Modulatoren direkt von den binären Ausgängen eines FPGA getrieben. Mit Treiberspannungen von lediglich 420 mV_{pp} werden Datenströme bis zu 52 Gbit/s erzeugt. Mit Treiberspannungen von 280 mV_{pp}, werden QPSK Signale mit einer Datenrate von 26 Gbit/s generiert und fehlerfrei über eine Strecke von 100 km übertragen.

Kapitel 6 fasst die Ergebnisse dieser Arbeit zusammen und bietet einen Ausblick auf künftige Aufgabenfelder und Herausforderungen in Bezug auf SOH-Modulatoren.

Preface

Communication is an indispensable part of toady's civilized life. The permanent availability of communication is based on a global network employing optical communication. Depending on the application scenarios, the technology and consequently the associated requirements differ vastly. This field faced enormous challenges to keep up with the growing demand for bandwidth and at the same time has undergone significant evolution in terms of technology and applications: data centers have become an indispensable part of the global network and this environment puts stringent power and energy requirements on the deployed optical communication modules in which electro-optic modulators are a key component. These requirements cannot be fulfilled by conventional lithium niobate ($LiNbO_3$) modulators as they are too bulky and cannot be compactly integrated. For this reason, semiconductor platforms have gained much attention in recent years. Leveraging processes from complementary metal–oxide–semiconductor (CMOS) technology, silicon is an attractive candidate because of its ability and potential for large-scale fabrication and mass production. Standard silicon phase modulators rely on the plasma dispersion effect and suffer from limited modulation efficiency and consequently pose limitations on energy-efficiency.

This thesis deals with an alternative approach for optical modulation on the silicon platform. Unlike conventional silicon modulators, the modulation mechanism relies on the second-order nonlinearity of an organic material which is dispensed onto the silicon chip. So-called silicon-organic hybrid (SOH) modulators outperform conventional silicon modulators by more than one order of magnitude in terms of the modulation efficiency. The fabrication process is compatible with standard optical lithography processes. In this thesis, SOH modulators are used to generate data signals at symbol rates up to 100 GBd – record values for semiconductor-based optical modulators. Furthermore, the challenge of energy efficient operation is addressed: SOH modulators are operated from the binary outputs of a field-programmable gate array (FPGA) without electrical driver amplifiers and with voltage levels below 300 mV_{pp}. Higher-order modulation formats are generated without the use of digital-to-analog converters (DAC) but simply through a passive radio-frequency (RF) network. Using drive voltages of a few hundred millivolts, the electrical energy consumption of the modulator amounts to less than 20 fJ/bit.

Chapter 1 gives an introduction to optical communications and provides insights into today's challenges regarding photonic integration.

Chapter 2 reviews the fundamentals of optical communications. In particular, optical signals including modulation formats and associated reception mechanisms are discussed. The second part covers electro-optic modulators and discusses the fundamentals of silicon-based modulators. In this context, the principle of SOH modulators is discussed in detail.

Chapter 3 addresses the signal transmission of on-off keying (OOK) signals using SOH Mach-Zehnder modulators (MZM). It describes the generation of 100 Gbit/s data signals with record-low drive voltages of only 1.4 V_{pp} resulting in an electrical energy consumption below 100 fJ/bit. It is shown that the high modulation efficiency and the resulting requirements of drive voltages lead to a mitigation of bandwidth limitations and allows generating optical data signals well beyond the bandwidth of the modulator and the associated RF circuitry.

Chapter 4 covers aspects of coherent communications using SOH in-phase/quadrature (IQ) modulators. Quadrature phase shift keying (QPSK) and 16-state quadrature amplitude modulation (16QAM) signals are generated at symbol rates up to 100 GBd. It was the first demonstration of a semiconductor-based optical modulator capable of 100 GBd 16QAM signal generation. To date, the resulting line rate of 400 Gbit/s is unmatched by competing device principles on the silicon platform.

Chapter 5 focusses on energy-efficient solutions for optical communications. It reports amplifier-less signal generation using SOH modulators. The SOH modulator is directly driven by the binary outputs of an FPGA, and data streams up to 52 Gbit/s are generated with drive voltages of only 420 mV_{pp}. At drive voltages of only 280 mV_{pp}, QPSK signal with a line rate of 26 Gbit/s are generated and transmitted without errors over a fiber span of 100 km.

Chapter 6 summarizes the work of this thesis. It further outlines next steps related to research and application of SOH modulators and SOH-based photonic integrated circuits (PIC).

Achievements of the present work

In this thesis, silicon-organic hybrid (SOH) modulators were investigated with a focus on high-speed and low-power optical data transmission for potential applications in telecom and datacom scenarios. While operating at the highest speed, SOH modulators simultaneously maintain a low energy consumption making them attractive candidates for application in energy-critical environments.

In the following, a detailed overview on the particular achievements is given:

First demonstration of 100 Gbit/s OOK for semiconductor-based modulators:

An SOH MZM was demonstrated for the generation of a 100 Gbit/s data stream using OOK modulation. The line rate and the electrical energy consumption of 98 fJ/bit are record values for semiconductor-based modulators generating OOK signals. The work represents the first demonstration of 100 Gbit/s OOK signaling on a semiconductor platform. In the meantime the line rate has been matched by other device concepts, but to date it still is the lowest energy consumption reported in literature for a semiconductor-based electro-optic modulator [J19], [C29], [C40].

First demonstration of 100 GBd 16QAM signaling with a semiconductor-based modulator:

SOH IQ modulators were demonstrated operating at symbol rates up to 100 GBd using 16QAM modulation. It is the first demonstration of 100 GBd operation for a semiconductor-based modulator for coherent applications. At the time of publication, the single-polarization line rate of 400 Gbit/s was the highest data rate generated with a semiconductor-based modulator which was to date only surpassed by indium-phosphide IQ modulators [J20], [C31], [C40], [C45].

First demonstration of amplifier-less operation of non-resonant silicon-based Mach-Zehnder modulators:

SOH modulators were driven directly from FPGA outputs without external drive amplifiers. Drive voltages are as low as 280 mV [J7], [C10], [C12], [C40].

DAC-less generation of advanced modulation formats:

The four-level drive signals for the I and Q-component of an SOH IQ MZM have been generated without the use of digital-to-analog converters (DAC). Instead, the signals were derived from the binary outputs of an FPGA and combined in passive radio frequency (RF) power combiners. A 52 Gbit/s data stream resulting from 16QAM modulation at 13 GBd was generated [J7], [C12].

Advanced modulation with record-low power consumption:

An IQ modulator based on the SOH technology was demonstrated with a record-low energy consumption of 18 fJ/bit for 13 GBd 16QAM modulation beating the previous record of another SOH IQ modulator demonstration [J7], [C12].

1 Introduction

In today's society, communications technology has become indispensable – the internet and mobile communications are equally essential for businesses and private persons. All that relies on the world-spanning fiber-optic network. Optical networks [1] have been made possible by groundbreaking scientific achievements which have been awarded with the Nobel Prize: in 1964, Charles Townes, Nikolay Basov and Aleksandr Prokhorov were honored with the Nobel Prize for their contributions in quantum electronics which led to the development of the laser [2]. 45 years later, in 2009, Charles Kao was awarded with the Nobel prize [3] for his "*groundbreaking achievements concerning the transmission of light in fibers for optical communication*". It is these two components – the light source and the transmission medium – that are at the heart of today's optical networks with millions of kilometers of fiber worldwide. The amount of data sent through this global network increases exponentially with time as projected and measured by Cisco and reported in its annual Networking Indices [4]. Other monumental achievements such as (external) waveguide modulators [5], wavelength-division multiplexing (WDM) [6], optical amplification, [7] and most recently coherent communication have made it possible to scale the network's capacity according to the demand, leading to communication links with capacities exceeding 1 Tbit/s.

The evolution of optical communication started in the long-haul domain and nowadays is also indispensable for short-reach applications. This is due to the changes in the content which is transmitted through the network: While in the early days of the internet, static webpages were accessed by the users, nowadays, dynamic websites are adapted according to the user's needs, and video streaming contributes a significant amount to the total data transmitted. To enable this technology, service providers like Google, Facebook and others operate warehouse-scale data centers around the world to host and process the data for their services. The largest portion of data is processed (and transmitted) internally and only a fraction is relevant for the end user. Data transmission in such data centers relies on optical communications as well. However, the requirements for short-reach links are strongly different from long-haul applications: The enormous density of optical interfaces poses limitations on their realization, such as size and power consumption. Consequently today, such links use intensity modulation and direct detection (IM/DD) rather than coherent communication. Conventional optical modulators realized in lithium niobate ($LiNbO_3$) [8] are rather bulky and not suited for large-scale integration. It is for that reason, that research on integrated optical modulators and eventually more complex photonic integrated circuits (PIC) has attracted much attention in recent years. Similar to Moore's law in the electronic domain, an exponential increase in the integration density of photonic components can be observed [9].

For the realization of compact optical modulators, a number of vastly different material platforms have been investigated and discussed. Similar to optical silica fibers, silica-on-silicon allows for low-loss silica waveguides. The platform offers easy fabrication, however, the footprint is comparatively large, and the technology misses optical nonlinearities for optical modulators nor does it offer light sources or photodetectors. All these elements can be realized on the indium phosphide (InP) platform [10] making it a complete photonic platform. The downside of InP is the fact that the material itself is expensive, and that the fabrication process involving epitaxial growth processes is complicated. The limited wafer size makes it less suitable for high volume mass production.

Leveraging the well-developed complementary metal–oxide–semiconductor (CMOS) technology, silicon (Si) photonics has emerged as a promising technology in recent years. As witnessed by fabrication of electronics, the silicon photonic platform is suited for large-scale production using large wafers of 12 inches or more and allows for small footprint PICs. Another advantage of the silicon platform is the availability of commercial foundry services [11]. Being an indirect semiconductor, silicon does not allow for light emission, but photodetectors have been realized in combination with germanium [12]. The inversion symmetry of silicon's crystalline structure prevents electro-optic modulators based on the Pockels effect. Instead, all-silicon photonic modulators [13] exploit the plasma-dispersion effect for phase modulation and a number of data transmission demonstrations have been reported [14]–[16].

The modulation efficiency of plasma-dispersion modulators, however, is limited and inferior to InP modulators by roughly a factor of two when expressed by the π-voltage-length product $U_\pi L$. Therefore, device lengths are typically in the order of several millimeters, otherwise the modulators require high drive voltages.

Another approach to realize optical phase modulators on the silicon platform is to combine the silicon photonic platform with organic electro-optic (EO) materials. It is referred to as silicon-organic hybrid (SOH) integration [17]. SOH modulators frequently use slotted waveguides where the slot is filled with an organic EO material [18]. The EO material provides a $\chi^{(2)}$-nonlinearity, and exploiting the linear electro-optic effect (Pockels effect), SOH modulators allow for pure phase modulation without residual amplitude modulation. Demonstration of SOH modulators show the enormous potential regarding bandwidth [19] and modulation efficiency [20] in terms of the π-voltage-length product $U_\pi L$ which outperforms conventional silicon modulators by roughly a factor of 20, and InP modulators by a factor of about 10. While the slot waveguide and its dimension in the order of only 100 nm is a critical aspect of SOH modulators, they can be fabricated in commercial optical lithography lines using, e.g., the 248 nm deep-ultraviolet (DUV) process [21]. Concerns have often been raised as to the stability of the organic material. To date, the operation at elevated temperatures was demonstrated [22], and further demonstrations of long-term stable materials are expected soon.

The work in this thesis addresses application aspects of SOH modulators for different application scenarios. High-speed signal generation is demonstrated for both coherent and IM/DD applications. Using a single Mach-Zehnder modulator (MZM), a 100 Gbit/s on-off keying (OOK) data stream was generated and transmitted over a dispersion-compensated link of 10 km SMF with a record-low drive voltage of only 1.4 V_{pp}. This was the first demonstration of 100 Gbit/s OOK signaling using a semiconductor-based modulator. It is shown and confirmed by simulation, that limitations of the modulator and driver bandwidth can be mitigated by exploiting the nonlinear transfer function of an MZM. This is enabled by the low π-voltage of only 0.9 V at direct current (DC). Using an in-phase/quadrature (IQ) modulator, a 400 Gbit/s data stream is generated at a symbol rate of 100 GBd with 16-state quadrature amplitude modulation (16QAM). It is the first demonstration that reaches the 100 GBd mark for coherent communication using semiconductor-based modulators. Looking into application scenarios for energy-efficient communication, an SOH modulator for sub-Volt drive signals is presented. It is directly operated from the binary interfaces of a field-programmable gate array (FPGA) to generate OOK and binary phase shift keying (BPSK) signals at 10 GBd. For the generation of 13 GBd quadrature phase shift keying (QPSK) and 16QAM signals, an IQ modulator is operated from an FPGA without amplifiers and digital-to-analog converters with a drive voltage as low as 280 mV_{pp} and 410 mV_{pp}, respectively. The experiments in this thesis – representing record results in terms of data rate and energy-efficiency for silicon-based modulators – demonstrate the enormous potential for SOH modulators for application in various sectors of optical communications.

2 Optical data signals and device concepts

An optical transmission system consists of various subsystems and components. A schematic overview of an optical transmission system is depicted in Figure 2.1. The overall transmission system can be divided into the transmitter, the channel and the receiver. The transmitter converts digital data from the electrical to the optical domain: Using digital signal processing (DSP), the digital bit stream is prepared and subsequently converted to the analog domain using a digital-to-analog converter (DAC). The electrical signal is then modulated onto an optical carrier. One discriminates into direct modulation and external modulation. In direct modulation, the laser current (and consequently the laser power) is directly modulated. This thesis focuses on external modulation where a continuous wave (CW) signal is modulated in a subsequent optical modulator.

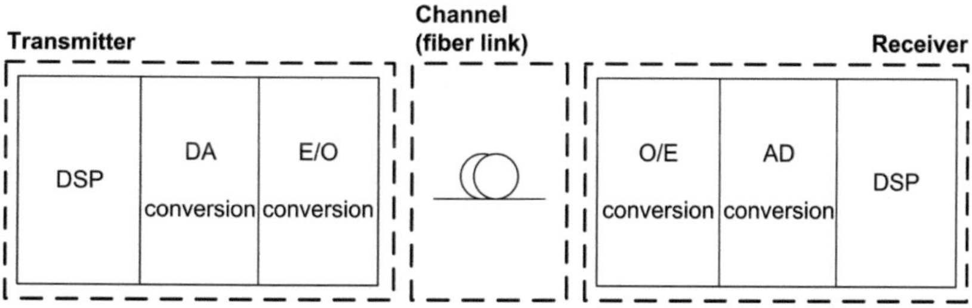

Figure 2.1: Block diagram of an optical communication system. In the transmitter, digital signal processing (DSP) prepares the digital signal to be transmitted over an optical link. After digital-to-analog (DA) conversion, the electrical signal is converted to an optical (electrical-to-optical (E/O) conversion). The signal is transmitted over the link, commonly an optical fiber link. At the receiver, the signal is converted to the electrical domain (optical-to-electrical (O/E) conversion) and digitized in an analog-to-digital (AD) converter. The subsequent DSP block extracts the digital data from the received signal.

Most commonly, optical data signals are transmitted in optical fibers which offer low losses and a tremendous bandwidth. Telecommunication and datacom applications most commonly rely on glass fibers (silica). Waveguiding in optical fibers bases on total internal reflection [23] and one fundamentally discriminates into single-mode fibers (SMF) in which only a single transverse mode can propagate, and multi-mode fibers (MMF) which allow the propagation of multiple modes. Alternative transmission media are plastic optical fibers (POF) [24] or free-space transmission [25]–[27]. Further, the reader may note that there is a multitude of specialty fiber such as multi-core fibers [28], nonlinear fibers [29], and photonic crystal fibers [30] which are not discussed here.

The reception of optical data signals relies on the conversion of the optical signal into the electrical domain. Subsequently, the signal is processed in a chain of DSP blocks in order to recover the digital data, see Appendix B.

In this chapter, we review key components of an optical data transmission system. In Section 2.1, optical signals are discussed. This includes optical modulation formats and pulse shapes as well as the corresponding reception schemes. Optical modulators, particularly on the silicon platform, are discussed in Section 2.2.

2.1 Optical signal generation and modulation formats

An optical signal is an electromagnetic wave. Mathematically, in the form of an analytic signal, we can describe the electric field *E(t)* by means of the optical carrier with angular frequency ω_0 and a "slowly varying" and generally complex amplitude $\underline{A}(t)$,

$$E(t) = \underline{A}(t)\exp(\mathrm{j}\omega_0 t). \tag{2.1}$$

It shall be noted that the physically measurable signal is real-valued and corresponds to the real-part of the above equation.

Digital data are transferred onto the optical carrier by means of modulation which can be understood as the manipulation of the optical carrier's physical properties. This section recalls the most common modulation types, namely intensity modulation and (coherent) quadrature amplitude modulation (QAM). Other modulation types such as frequency modulation [31], polarization modulation [32], pulse position modulation (PPM) [33] or combinations thereof [34] are not discussed here. Further in this section, we discuss the corresponding reception mechanisms, namely, direct detection and coherent detection, and we review quality metrics for these modulation formats.

Following the notation in [35], we can express a digital signal by a discrete set of symbols a_k which in general may be complex with a real and imaginary part, $a_k = a_k^r + \mathrm{j} a_k^i$. Presuming equal probability of all symbols, one encodes m bits in $M = 2^m$ symbols with a symbol duration T_s. The bit rate (or line rate) r of the signal is related to the symbol rate $R_S = 1/T_s$ and the number of symbols in the set (or bits),

$$r = R_S \log_2 M = R_S m. \tag{2.2}$$

The unit of the symbol rate is *Baud* (Bd) which is defined as 1 Bd = 1 symbol/s. For $M = 2$, one refers to binary modulation. With two states, one bit per symbol can be encoded, and the data rate equals the symbol rate. For $M > 2$ ($m > 1$) and at a constant symbol rate, one transmits more data on the carrier. This relates to the spectral efficiency (SE) which is a quantity of how much data are transmitted per time and per bandwidth. The spectral efficiency is given in units of bit/s/Hz.

A time-continuous data signal can then be described by the symbol $a_{k(i)}$ in the time interval (i), by a modulation waveform $p(t)$ and by the symbol duration T_s,

$$x(t) = \sum_{i=-\infty}^{\infty} a_{k(i)} p(t - iT_S). \tag{2.3}$$

The modulation waveform $p(t)$ may contain chirping or pulse shaping (see section 2.1.3 for details).

2.1.1 Intensity modulation and direct detection

Traditionally, optical transmission systems relied on on-off keying (OOK) as the typical modulation format. In this type of binary modulation, the optical data signal is switched between "On" and "Off", relating to an alphabet $a_k = \{0,1\}$. The reception process is simple: The optical power is detected by a photodiode and converted to an electrical signal. The photocurrent is linearly related to the received power $P \sim |E|^2$, and to the sensitivity (responsivity) of the photodiode is denoted by S,

$$i(t) = SP(t). \tag{2.4}$$

Typically, the photocurrent $i(t)$ is converted into an electrical voltage $u(t)$ either by a simple resistor or by means of a transimpedance amplifier (TIA). The sensitivity describes the photodiode's conversion efficiency and is related to the elementary charge e, the quantum efficiency η, the reduced Planck constant $\hbar$ and the angular frequency ω_0 of the incident light,

$$S = \frac{e\eta}{\hbar\omega_0}. \tag{2.5}$$

For communication purposes, the bit error ratio (BER) is the typical quality metric. On-off keying signals can also be evaluated by a few other figures of merit. Visually, it is characterized by an eye diagram, Figure 2.2. The eye diagram illustrates the shape of the signal in the time domain superimposed a certain number of symbols. Quantitative measures are the extinction ratio (ER) and the quality factor (Q-factor). The extinction ratio describes how well the power in the "Off"-state is suppressed referred to the "On"-state. The Q-factor is a quantity which relates the mean electrical levels u_0, u_1 and the corresponding standard deviations σ_0, σ_1 of the "Off" and "On"-state. It can be related to a BER [36]:

$$Q = \frac{u_1 - u_0}{\sigma_1 + \sigma_0}, \tag{2.6}$$

$$\mathrm{BER} = \frac{1}{2}\mathrm{erfc}\left(\frac{Q}{\sqrt{2}}\right). \tag{2.7}$$

The Q-factor is especially helpful to analyze or estimate signal quality if real-time analysis of the BER is not possible or if data recordings are not long enough for statistically relevant BER analysis. The relation of Q and BER uses the complementary error function

$$\mathrm{erfc}(z) = \frac{2}{\sqrt{\pi}}\int_z^{\infty} \exp\left(-t^2\right)\mathrm{d}t \tag{2.8}$$

and is valid under the assumptions that the sampling time and the decision threshold were chosen optimally, that "zeros" and "ones" appear with equal probability, that inter-symbol interference is excluded, and the noise follows a Gaussian probability distribution. A detailed description of the conversion of the Q-factor to a BER can be found in Appendix C. In general, Eq. (2.7) cannot be applied for optically pre-amplified signals for which the signal quality is limited by optical noise: Due to the power-law detection, the assumption of Gaussian noise does not hold any more for the electrical signal.

More recently, intensity-modulated signals with more than two intensity levels have attracted quite some attention [37]–[41]. So-called pulse-amplitude modulation (PAM) signals encode more bits per symbol, allowing a higher spectral efficiency at the expense of reduced noise resilience.

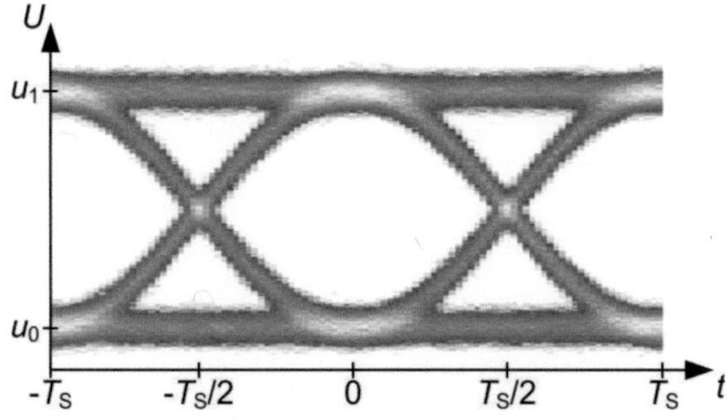

Figure 2.2: Eye diagram of an on-off keying (OOK) signal. The eye diagram is a visual quality metric of an OOK signal – a wide open "eye" represents a good signal quality.

2.1.2 Coherent modulation formats

In intensity modulation, any phase information of the optical carrier is disregarded. However, we can also include the phase information, e.g., by defining alphabets $a_k = \{-1,+1\}$ or $a_k = \{-3,-1,+1,+3\}$, typically referred to as binary phase shift keying (BPSK) and 4-level bipolar amplitude shift keying (4ASK), respectively. The so-called constellation diagrams for BPSK and 4ASK modulation are shown in Figure 2.3 (a) and (b). These modulation formats still only use the real part of an optical signal, but modulation formats can be extended into the complex plane: complex modulation formats use both the real and imaginary part – referred to in-phase (I) and quadrature (Q) component – to carry information and are nowadays widely used, both, in academia and industry [42]–[44]. These modulation formats can be limited to phase modulation (phase shift keying, PSK) only, $|a_k| = 1$, such as quadrature phase shift keying (QPSK) with 4 constellation points, $a_k = \{-1+\mathrm{j}, -1-\mathrm{j}, +1+\mathrm{j}, +1-\mathrm{j}\}$, where j denotes the imaginary unit, see Figure 2.3 (c), or higher-order phase modulation formats such as 8-state PSK (8PSK, not depicted). Another typically used modulation format is quadrature amplitude modulation (QAM), e.g., the constellation diagram for 16-state QAM (16QAM) is depicted in Figure 2.3 (d). Higher-order QAM signals are not depicted in the figure, but 32QAM and 64QAM as well as 512QAM and 1024QAM have been demonstrated in a number of experiments [14], [45]–[47] . In general, the alphabets for square-shaped QAM signals are given as

$$
\begin{aligned}
a_k &= a_k^r + \mathrm{j} a_k^i \\
a_k^r, a_k^i &= \left\{ \begin{array}{l} -\left(\sqrt{M}-1\right), -\left(\sqrt{M}-3\right), \ldots, -3, -1, \\ \qquad +1, +3, \ldots, +\left(\sqrt{M}-3\right), +\left(\sqrt{M}-1\right) \end{array} \right\}
\end{aligned}
\tag{2.9}
$$

Unlike QPSK, 16QAM, 64QAM and others, some constellations are not arranged in a square but leave out some constellation points in the corners, e.g., 32QAM, 128QAM and 512QAM. As a consequence, the equation above cannot be used to define the symbols for these modulation formats.

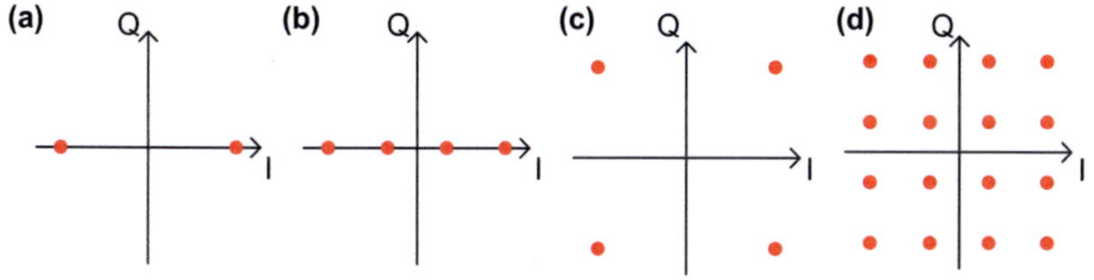

Figure 2.3: Coherent modulation formats. Coherent modulation formats can either have only real-valued symbols, as for **(a)** binary phase-shift keying (BPSK) and **(b)** 4-level bipolar amplitude-shift keying (4ASK), or use complex-valued symbols such as **(c)** quadrature phase shift keying (QPSK) and **(d)** 16-state quadrature amplitude modulation (16QAM).

Similar to the case of PAM, the spectral efficiency and the data rate increase with the order of the modulation format, but the signal requires a higher signal-to-noise power ratio (SNR) for the same bit error ratio (BER) [48].

2.1.2.1 Coherent detection

In the previous section, it was stated that direct detection only converts the optical power into the electrical domain but the phase information is lost. The reception of bipolar ASK, PSK and QAM signals therefore relies on phase-sensitive coherent reception. This section follows the derivation in [49]. For coherent reception, we superimpose the optical signal

$$E_{\mathrm{sig}} = A_{\mathrm{sig}}(t)\exp\left(\mathrm{j}\left(\omega_0 t + \varphi_{\mathrm{sig}}(t)\right)\right) \tag{2.10}$$

with a so-called local oscillator (LO) at an angular frequency ω_{LO}

$$E_{\mathrm{LO}} = A_{\mathrm{LO}}\exp\left(\mathrm{j}\left(\omega_{\mathrm{LO}} t + \varphi_{\mathrm{LO}}\right)\right). \tag{2.11}$$

It is to be noted, that in Eqs. (2.10) and (2.11), the complex amplitude $\underline{A}$ from in Eq. (2.1) was separated into a real-valued amplitude term A and a phase term φ included in the complex exponential function. Further, we omit the time-dependence for the amplitude and phase of the LO such that we neglect amplitude and phase noise of the LO laser.

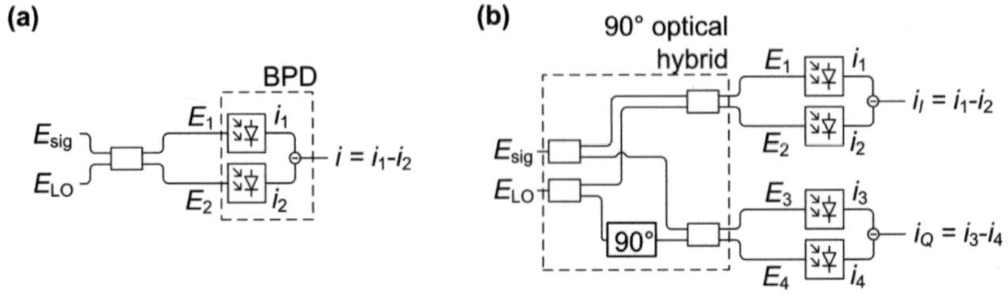

Figure 2.4: Coherent receivers. (a) Phase information of the optical signal can be recovered through the beating. Therefore, the optical signal E_{sig} is superimposed with a local oscillator (LO) E_{LO} using an optical 2x2 coupler. The two outputs are coupled to a balanced photodetector (BPD) consisting of two photodiodes of which the two photocurrents are subtracted from each other. (b) Coherent receiver for complex-valued modulation formats. Amplitude and phase information of the signal can be recovered by combing two receivers as depicted in (a) and introducing a 90° phase shift for the LO in one arm.

As depicted in Figure 2.4 (a), the two fields are superimposed in an optical 3 dB coupler with a 50:50 coupling ratio which adds a 180° phase shift to the signal or to the oscillator between the two coupler outputs. Subsequently, the superimposed signals are detected on two photodiodes at both outputs of the optical coupler. Provided that the two fields share the same state of polarization (SOP), the fields at the two outputs of the coupler are [49]

$$E_1(t) = \frac{1}{\sqrt{2}}\left(E_{\text{sig}}(t) + E_{\text{LO}}(t)\right), \tag{2.12}$$

$$E_2(t) = \frac{1}{\sqrt{2}}\left(E_{\text{sig}}(t) - E_{\text{LO}}(t)\right). \tag{2.13}$$

The beating of signal and LO leads to photocurrents

$$i_1(t) = SP_1 \sim \frac{1}{2}\left|E_{\text{sig}}\right|^2 + \frac{1}{2}\left|E_{\text{LO}}\right|^2 + \left|E_{\text{sig}}\right|\left|E_{\text{LO}}\right|\cos\left(\Delta\omega t + \Delta\varphi(t)\right), \tag{2.14}$$

$$i_2(t) = SP_2 \sim \frac{1}{2}\left|E_{\text{sig}}\right|^2 + \frac{1}{2}\left|E_{\text{LO}}\right|^2 - \left|E_{\text{sig}}\right|\left|E_{\text{LO}}\right|\cos\left(\Delta\omega t + \Delta\varphi(t)\right), \tag{2.15}$$

where $\Delta\omega = \omega_{\text{sig}} - \omega_{\text{LO}}$ is the difference frequency and $\Delta\varphi(t) = \varphi_{\text{sig}}(t) - \varphi_{\text{LO}}$ is the phase difference of the signal and the local oscillator. Using a balanced detector in which the photocurrents of the two photodiodes are subtracted, then yields a combined photocurrent

$$i(t) = i_1(t) - i_2(t) \sim \left|E_{\text{sig}}\right|\left|E_{\text{LO}}\right| \cos\left(\Delta\omega t + \Delta\varphi(t)\right). \tag{2.16}$$

The DC terms $\left|E_{\text{sig}}\right|^2$ and $\left|E_{\text{LO}}\right|^2$ have vanished, leaving only the mixing term which is of interest. From this beat signal, however, we cannot yet derive the complex field amplitude. Eq. (2.16) only gives a measure of the amplitude component oriented in one dimension of the complex plane.

The orthogonal dimension can be retrieved if a second path is introduced in which the local oscillator is phase shifted by $\pi/2$. To this end, a so-called 90°-hybrid is used, see Figure 2.4 (b). It extends the coherent receiver shown in Figure 2.4 (a) by a second arm with a 90° phase shift in the LO path. In this structure, the electric fields at the four optical outputs can be summarized as

$$E_1(t) = \frac{1}{2}\left(E_{\text{sig}}(t) + E_{\text{LO}}(t)\right) \tag{2.17}$$

$$E_2(t) = \frac{1}{2}\left(E_{\text{sig}}(t) - E_{\text{LO}}(t)\right). \tag{2.18}$$

$$E_3(t) = \frac{1}{2}\left(E_{\text{sig}}(t) + \mathrm{j}E_{\text{LO}}(t)\right) \tag{2.19}$$

$$E_4(t) = \frac{1}{2}\left(E_{\text{sig}}(t) - \mathrm{j}E_{\text{LO}}(t)\right). \tag{2.20}$$

The photocurrents i_I and i_Q, of the two balanced detectors, are related to the sine and cosine of the instantaneous optical phase differences and therefore refer to the in-phase and quadrature component:

$$i_I(t) = i_1(t) - i_2(t) \sim \left|E_{\text{sig}}\right|\left|E_{\text{LO}}\right| \cos\left(\Delta\omega t + \Delta\varphi(t)\right) \tag{2.21}$$

$$i_Q(t) = i_3(t) - i_4(t) \sim \left|E_{\text{sig}}\right|\left|E_{\text{LO}}\right| \sin\left(\Delta\omega t + \Delta\varphi(t)\right) \tag{2.22}$$

Using i_I and i_Q, the complex-valued (analytic) signal can be reconstructed [49]

$$i_C(t) = i_I(t) + \mathrm{j}i_Q(t) \sim |E_{\mathrm{sig}}||E_{\mathrm{LO}}|\exp\left(\mathrm{j}\left(\Delta\omega t + \Delta\varphi(t)\right)\right). \tag{2.23}$$

Another essential part in coherent communication is the decoding and compensation of signal impairments. Such impairments comprise, e.g., phase (and amplitude) noise of the transmitter laser and local oscillator, a frequency offset of both lasers, and transmission effects such as chromatic dispersion and polarization rotation. These effects are typically compensated digitally. A short overview on the major DSP blocks is summarized in Appendix B.

The reception mechanism discussed in this section can be used for coherent modulation formats in general. However, there are distinct modulation and coding schemes in which the receiver circuit can be simplified. Such differential detection schemes can be applied for PSK (and QPSK) modulation, which are then referred to as differential phase shift keying (DPSK) and differential quadrature amplitude modulation (DQPSK). Differential receivers use delay-interferometers to superimpose the signal with a 1 symbol-time-delayed copy of itself. Thereby, the phase of the signal is evaluated with respect to the preceding symbol instead of to an extra reference. Differential reception is not discussed in detail here. More detailed information on differential applications can be found in [50].

2.1.2.2 Error vector magnitude as a quality metric for coherent modulation

The Q-factor as defined in Eq. (2.6) is only defined for a binary signal and cannot be used for complex modulation formats. Instead, the error vector magnitude (EVM) [51], [52] can be used. In the previous section, we defined the ideal constellation points for QAM signals. The EVM now quantifies how far a received symbol is from its ideal constellation point, see Figure 2.5. We define EVM as the root mean square of an error vector E_{err} of N randomly transmitted symbols related to the maximum vector of the ideal constellation diagram $E_{\mathrm{t,\,max}}$ [51], [52]

$$\mathrm{EVM_m} = \sqrt{\frac{1}{N}\sum_{i=1}^{N}\left|E_{\mathrm{err},i}\right|^2 \Big/ \left|E_{\mathrm{t,max}}\right|^2} \tag{2.24}$$

The definition above includes all signal distortions (also nonlinear ones) and uses the longest vector of a constellation for normalization which is indicated by the subscript 'm'. It is to be noted that another definition of the EVM normalizes to the average power of the M constellation points. Both definitions are related by a correction factor k_{norm},

$$k_{\mathrm{norm}}^2 = \left|E_{\mathrm{t,max}}\right|^2 \Big/ \frac{1}{M}\sum_{i=1}^{M}\left|E_{\mathrm{t},i}\right|^2, \tag{2.25}$$

which is dependent on the modulation format [51], [52]. Using the number L_{N} of signal levels identical within each dimension of the (quadratic) constellation, one can convert the $\mathrm{EVM_m}$ to a BER according to [51], [52]

$$\mathrm{BER} \approx \frac{\left(1-L_{\mathrm{N}}^{-1}\right)}{\log_2 L_{\mathrm{N}}}\,\mathrm{erfc}\left(\sqrt{\frac{3\log_2 L_{\mathrm{N}}}{L_N^2-1}\,\frac{1}{\left(k_{\mathrm{norm}}\mathrm{EVM_m}\right)^2\log_2 M}}\right). \tag{2.26}$$

The complementary error function erfc is defined in Eq. (2.8). Similar to the conversion from Q-factor to BER, the conversion in Eq. (2.26) assumes Gaussian noise. More details on EVM are found in reference [51], [52].

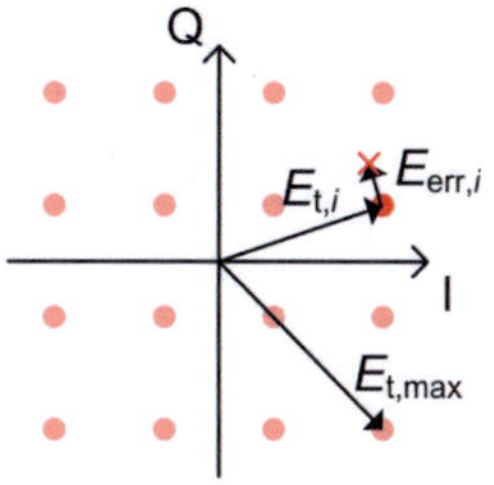

Figure 2.5: Error vector magnitude (EVM). The error vector magnitude is a quality metric for complex modulation formats. It relates the deviation $E_{\mathrm{err},i}$ of a measured constellation point from its ideal position $E_{\mathrm{t},i}$. The definition of $\mathrm{EVM_m}$ is normalized to the maximum vector length $E_{\mathrm{t},\,max}$ of the modulation format. EVM can be related to a BER under the assumption of Gaussian noise limitation.

2.1.3 Pulse shaping for optical communications

Spectral efficiency – as stated above – depends on the number of bits encoded per symbol (choice of modulation format). Particularly in wavelength-division multiplexing (WDM) systems, the channel spacing is another quantity that affects SE. The channel spacing in turn is dictated by the modulated carrier's spectral shape and by potentially required guard bands between the carriers.

Conventionally, optical transmission systems used non-return-to-zero (NRZ) modulation with an (ideally) rectangular pulse shape in time domain. The spectral shape is given by the Fourier transform of the time-domain pulse shape. Hence, for a rectangular pulse shape, the spectrum is sinc-shaped and features sidelobes. While these sidelobes would theoretically extend until infinity, in practical scenarios they are suppressed by a bandwidth limitation of the electro-optic components. Nevertheless, to avoid inter-channel interference (ICI), a guard band has to be inserted. Such a guard band, however, effectively represents unused bandwidth and consequently limits the achievable spectral efficiency. Alternatively, the spectral spread of a channel can be limited by adapting the time-domain pulse shape. In the limiting case, the spectral shape can be confined to a rectangular power spectrum which is half as broad as the main lobe of a channel with rectangular pulses and equal symbol rate, see Figure 2.6 (a).

A commonly used pulse-shaped signal referred to as 'raised-cosine' (RC) signal exhibits an amplitude spectrum $\tilde{p}_{\mathrm{RC}}(f)$ with raised-cosine shape according to [53]

$$\tilde{p}_{\mathrm{RC}}(f)=\begin{cases} T_{\mathrm{S}}, & |f|\le\dfrac{1-\beta}{2T_{\mathrm{S}}} \\ T_{\mathrm{S}}\left\{1+\cos\left(\dfrac{\pi T_{\mathrm{S}}}{\beta}\left(|f|-\dfrac{1-\beta}{2T_{\mathrm{S}}}\right)\right)\right\}, & \dfrac{1-\beta}{2T_{\mathrm{S}}}<|f|<\dfrac{1+\beta}{2T_{\mathrm{S}}}, \\ 0, & \dfrac{1+\beta}{2T_{\mathrm{S}}}\le|f| \end{cases} \tag{2.27}$$

where β is referred to as the roll-off factor and takes values between zero and one. The roll-off factor β describes the excess bandwidth beyond the signal's Nyquist frequency $1/2T_S$, e.g., for $\beta = 1/2$ the excess bandwidth is 50 %. The corresponding time-domain pulse shape $p_{RC}(t)$ of an RC signal is [54]

$$p_{RC}(t) = \frac{\sin(\pi t/T_S)}{\pi t/T_S} \frac{\cos(\pi \beta t/T_S)}{1 - 4\beta^2 t^2/T_S^{\ 2}}. \tag{2.28}$$

From the equation above, we find that the sinc-pulse is a special case of the RC pulse shape with $\beta = 0$. For $\beta \neq 0$, the time domain pulse shape is a 'sinc-like function'.

Unlike a rectangular pulse, the pulses of an RC signal are not confined to the duration of a single symbol T_S but expand further into neighboring time slots. Nevertheless, pulses of RC signals fulfill the condition [54]

$$p_{RC}(iT_S) = \begin{cases} 1, & \text{for } i = 0 \\ 0, & \text{for } i = \pm 1, \pm 2, \ldots \end{cases}, \tag{2.29}$$

stating that each pulse is zero in the center of any neighboring pulse, Figure 2.6 (b). Fulfilling the Nyquist criterion, the pulse shape of an RC signal belongs to the group of Nyquist pulses, specifying that they are free from intersymbol interference (ISI).

Another commonly used pulse shape exhibits a spectrum shaped as a (square) root-raised cosine (RRC). RRC signals defined by [53]

$$\left|\tilde{p}_{RRC}(f)\right| = \sqrt{\left|\tilde{p}_{RC}(f)\right|}, \tag{2.30}$$

are not free from ISI. Instead, their appeal results from matched filter theory: when using an RRC-shaped receiver (Rx) filter, it acts as a matched filter on the signal transmitted, hence maximizing the signal's SNR [53]. At the same time and according to Eq. (2.30), applying an RRC Rx filter on an RRC signal yields an RC signal, and hence removes ISI. Thereby, RRC signals combine the limited spectral width of pulse-shaped signals with a matched filter maximizing the signal's SNR.

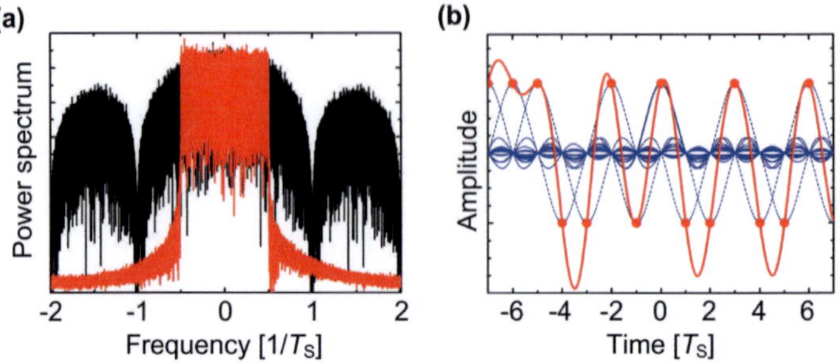

Figure 2.6: Pulse shaping in optical communications. **(a)** The spectrum of conventional NRZ signals (black) features sidelobes leading to a broad sprectral width of a data signal. The spectral shape can be narrowed down such that several data channels can be spaced more closely. The spectrum of a 'raised cosine' (RC) signal with roll-off $\beta = 0$ (red) has a rectangular shape. **(b)** The time-domain pulse shape of an RC signal with $\beta = 0$ is a sinc-pulse (solid) and fulfills the Nyquist criterion, stating that neighboring pulses (dashed) are zero in the center of any other symbol. The combined signal (red) is hence free from inter-symbol (ISI) interference.

More information on the real-time generation of pulse-shaped signals and corresponding aspects such as latency and computational complexity can be found in [55] and [56].

2.2 Principle of Mach-Zehnder and IQ modulators

The optical modulator is a key component of the optical transmitter subsystem. This section limits itself to Mach-Zehnder modulators (MZM); other modulator types such as resonant devices [57], [58], electro-absorption modulators (EAM) [59] as well as direct modulation of lasers are excluded here. In the first part we will review the function principle of a MZM. In the following subsections we will then discuss the realization of modulators on the silicon platform. The principle of an MZM bases on the Mach-Zehnder interferometer: an input electric field $E_{in}(t)$ is split into equal parts in an optical power divider. Phase shifter sections in both arms of the MZM allow manipulating the optical signal before recombination in a second optical combiner, Figure 2.7.

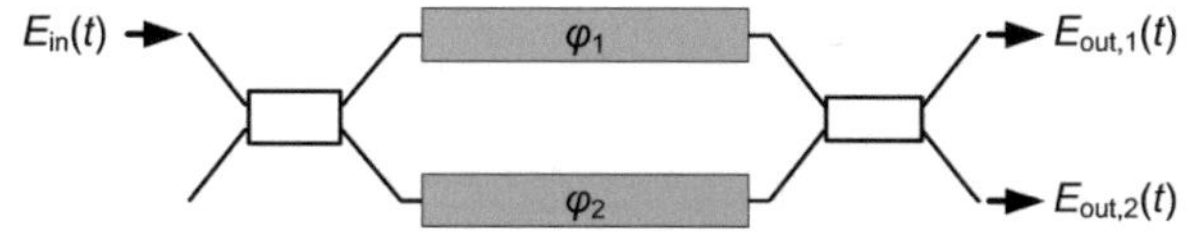

Figure 2.7: Schematic of a Mach-Zehnder modulator (MZM). The incoming optical signal $E_{in}(t)$ is split in two arms, in which the phase can be modulated. Subsequently, the light is combined in an output coupler and the signals from the two arms interfere to form the output signal.

The optical combiner and divider can be realized as multi-mode interference (MMI) couplers. For a 2 × 2 MMI coupler, we can write the two input and two output fields as vectors, $\vec{E}_{in}$, $\vec{E}_{out}$ and the transfer function of a (lossless) MMI coupler can be described by a matrix

$$\mathbf{M}_{2\times 2} = \frac{1}{\sqrt{2}}\begin{pmatrix} 1 & -j \\ -j & 1 \end{pmatrix}. \tag{2.31}$$

The electric fields of the optical signal at the two outputs of the MZM, $E_{out,1}(t)$ and $E_{out,2}(t)$, can then be described by the input electric field $E_{in}(t)$ and by the phase shifts in the two arms of the MZM, φ_1 and φ_2:

$$\begin{aligned} E_{out,1}(t) &= jE_{in}(t)\exp\left(j\frac{\varphi_1+\varphi_2}{2}\right)\cos\left(\frac{\varphi_1-\varphi_2}{2}\right) \\ E_{out,2}(t) &= jE_{in}(t)\exp\left(j\frac{\varphi_1+\varphi_2}{2}\right)\sin\left(\frac{\varphi_1-\varphi_2}{2}\right) \end{aligned} \tag{2.32}$$

In the above equation, the time-dependence of the phases $\varphi_1(t)$ and $\varphi_2(t)$ during modulation was omitted in favor of improved readability. Modulating the phases $\varphi_1(t)$ and $\varphi_2(t)$ in both arms of the MZM refers to a so-called dual-drive MZM. An MZM has two outputs. In practice, however, only one port is connected. The second port may be used as a monitor port for automatic bias control systems. Therefore, in the following considerations, only one output $E_{out}(t)$ will be considered. Eq. (2.32) describes the general transfer function for an MZM and can be simplified for distinct operation modes: in push-push operation, the time-dependent phase shifts in the two MZM arms are equal, $\varphi(t) = \varphi_1(t) = \varphi_2(t)$, and the MZM acts as a phase modulator:

$$E_{\text{out}}(t) = \mathrm{j}E_{\text{in}}(t)\exp(\mathrm{j}\varphi(t)) \tag{2.33}$$

In push-pull operation, the phase shifts of the two arms are equal but with opposite sign, $\varphi(t) = \varphi_1(t) = -\varphi_2(t)$. Then, the MZM acts as a pure amplitude modulator without residual phase modulation:

$$E_{\text{out}}(t) = \mathrm{j}E_{\text{in}}(t)\cos(\varphi(t) + \varphi_{\text{bias}}) \tag{2.34}$$

In Eq. (2.34), we include a biasing phase shift, φ_{bias}, which can be used to adjust the operating point of the modulator. Typically, for amplitude modulation, the MZM is biased in the null point, $\varphi_{\text{bias}} = -\pi/2$. For intensity modulation, the modulator is operated in the quadrature (or 3 dB) point, $\varphi_{\text{bias}} = -\pi/4$. The (amplitude) transfer function for push-pull operation of a MZM, Eq. (2.34), can be re-written in order to give the intensity transfer function as follows

$$I_{\text{out}} = I_{\text{in}}\left(\frac{1}{2} + \frac{1}{2}\cos(2\varphi(t) + 2\varphi_{\text{bias}})\right). \tag{2.35}$$

Push-push and push-pull operation only require a single signal to operate the device. Unless further noted, we assume an MZM operated in push-pull mode in the following. While in principle and according to Eq. (2.32) any point in the complex plain can be addressed by a dual-drive MZM, this is not possible in push-pull operation. In order to address the complex plane, an in-phase/ quadrature (IQ) modulator consisting of two nested MZM can be used. In this configuration, two "child" MZM are arranged in a "parent" MZM, Figure 2.8. One MZM modulates the real part, the so-called in-phase (I) component. An additional phase shifter introduces a 90°-phase shift such that the signal of the second arm of the IQ structure can be interpreted as the imaginary part or quadrature (Q) component. Using an IQ modulator rather than a dual-drive MZM allows to use simple and independent drive signals for the I and Q component. The analytic signal at the output of the IQ modulator can be described by the phase shifts φ_{I} and φ_{Q} in the two MZM:

$$E_{\text{out}}(t) = E_{\text{in}}(t)\left(\frac{1}{2}\sin(\varphi_{\text{I}}(t)) + \mathrm{j}\frac{1}{2}\sin(\varphi_{\text{Q}}(t))\right) \tag{2.36}$$

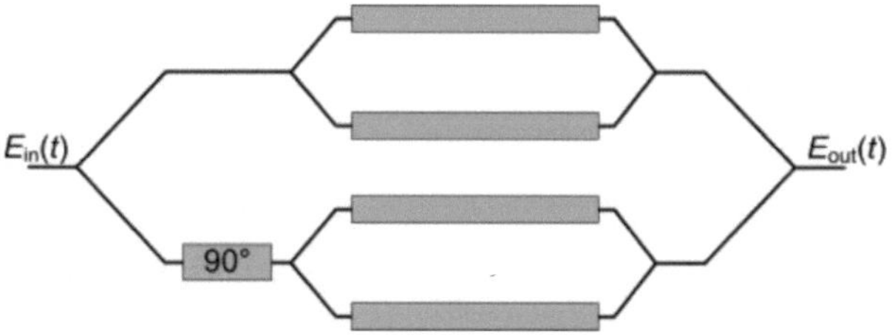

Figure 2.8: Schematic of an in-phase/quadrature (IQ) modulator. Two "child" MZM are arranged in a "parent" MZM. In-phase and quadrature component are modulated independently with push-pull MZM. A static 90°-phase shift allows to address the complex plane.

Note that in the equation above, the bias terms have been omitted. Instead, the typical operating point at null transmission is implicitly included in the sine notation.

Conventional electro-optic modulators, e.g., realized on lithium niobate ($LiNbO_3$), exploit the linear electro-optic effect, also referred to Pockels effect. A summary of the Pockels effect can be found in Appendix A and in Reference [60]. A more detailed discussion of optical nonlinearities can be found in Reference [61]. In essence, the linear electro-optic effect leads to an optical phase shift that is linearly dependent on a modulating voltage. In this context, a common figure of merit for an electro-optic phase shifter is the π-voltage $U_{\pi,\mathrm{PM}}$ for which the optical phase is shifted by $\varphi_{\mathrm{PM}} = \pi$. Similarly, a π-voltage $U_{\pi,\mathrm{MZM}}$ can be defined for a push-pull Mach-Zehnder structure with two parallel phase shifters. It is commonly defined by a relative phase difference of π in the two arms, $\Delta\varphi_{\mathrm{MZM}} = \varphi_1\text{-}\varphi_2 = 2\varphi = \pi$. In other words, we define the π-voltage for a push-pull MZM as the voltage that is required to switch between full transmission (constructive interference, "bright") and zero transmission (destructive interference, "dark"). Unless otherwise noted, we use U_π for the π-voltage of an MZM and omit the subscript 'MZM'. Similar to the modulating voltage, also the phase shifter length linearly affects the phase shift, and consequently, the π-voltage-length product $U_\pi L$ is another figure of merit that describes the modulation efficiency of an electro-optic modulator.

Using the π-voltage, we can also express the phase shifts by a modulating voltage $u(t)$ and the π-voltage U_π and rewrite the amplitude and intensity transfer functions, Eqs. (2.34) and (2.35), and the transfer function of an IQ modulator, Eq. (2.36) ,

$$E_{\text{out}}(t) = \text{j}E_{\text{in}}(t)\cos\left(\frac{u(t)}{U_\pi}\frac{\pi}{2}\right) \tag{2.37}$$

$$I_{\text{out}} = I_{\text{in}}\left(\frac{1}{2} + \frac{1}{2}\cos\left(\frac{u(t)}{U_\pi}\pi\right)\right) \tag{2.38}$$

$$E_{\text{out}}(t) = E_{\text{in}}(t)\left(\frac{1}{2}\sin\left(\frac{u_{\text{I}}(t)}{U_{\pi,\text{I}}}\frac{\pi}{2}\right) + \text{j}\frac{1}{2}\sin\left(\frac{u_{\text{Q}}(t)}{U_{\pi,\text{Q}}}\frac{\pi}{2}\right)\right) \tag{2.39}$$

Note that in practice, the π-voltages of the two "child" MZM in the IQ modulator might differ (slightly) such that we discriminate between $U_{\pi,\text{I}}$ and $U_{\pi,\text{Q}}$.

2.2.1 All-silicon electro-optic modulators

For a long time, $LiNbO_3$ modulators exploiting the Pockels effect have been the most commonly used modulators. Such modulators, however, are bulky with lengths of several centimeters and not suited for dense photonic integration. This gap in the photonic portfolio can be filled by semiconductor-based optical modulators. Offering the potential for co-integration with electronics [62], this section will focus on silicon-photonic modulators. While this thesis covers silicon-organic hybrid (SOH) modulators, conventional, all-silicon modulators are introduced here and serve as a benchmark for SOH devices. Other material platforms such as indium phosphide (InP) are not discussed here.

On the silicon platform, the electro-optic effect does not exist [63], because of the inversion symmetry of the crystalline silicon. Instead, all-silicon modulators typically rely on the plasma-dispersion effect which changes the refractive index of the material in dependence of the concentration of free carriers [63]. Hence, a change in the absorption also modulates the amplitude of the optical signal [63]. One discriminates three types of silicon photonic modulators [13]: carrier injection, carrier accumulation, and carrier depletion, see Figure 2.9.

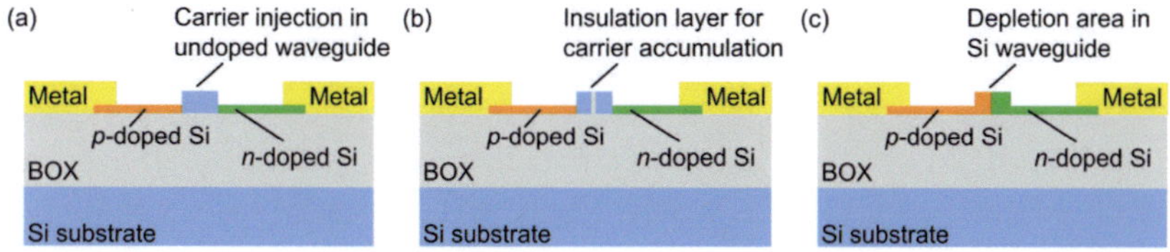

Figure 2.9: Schematic of different types silicon photonic modulators based on the plasma dispersion effect. (a) Carrier injection modulator. The waveguide is realized in the intrinsic section which is connected to the metal contacts via p and n-doped sections. (b) Carrier accumulation modulator. An additional insulating layer in the intrinsic section leads to the accumulation of carriers and avoids a current flow. (c) Carrier depletion modulator. Waveguide is integrated in a p-n junction. Unlike the injection and accumulation-type modulators, depletion-type modulators are operated under reverse bias.

For carrier injection modulators, the photonic waveguide is realized in the intrinsic section of a p-i-n junction that is operated under forward bias. The associated current flow consequently leads to comparatively high energy consumption. Carrier accumulation modulators feature a similar structure but have an additional insulation layer in the waveguide. While carrier injection [64], [65] and accumulation layer [16], [66] modulators feature $U_{\pi}L$-product around 1 Vmm, the bandwidth is limited by a slow minority recombination dynamics and an intrinsic capacitance, respectively [13], [63], [64]. In contrast, carrier depletion modulators with a p-n junction in the waveguide layer are operated in reverse bias and allow for high modulation speeds, but offer only a limited modulation efficiency around 10 Vmm [67], [68]. As a consequence, high-speed depletion-type silicon modulators require high drive voltages even for modulator lengths of several millimeters and hence, the associated energy per bit is in the order of 1 pJ/bit. This number can be hardly optimized as the phase shift is induced by a depletion (or injection) of carriers. The energy consumption hence is given by the required amount of carriers that is moved and the modulator design may merely trade drive voltage for drive current [17].

2.2.2 Silicon-organic hybrid (SOH) modulators

An alternative approach on the silicon platform is the so-called silicon-organic hybrid (SOH) technology which combines the silicon photonic platform with organic electro-optic (EO) materials. Thus, SOH integration maintains the advantages of silicon photonics such as the small footprint and mature processing and adds the second-order nonlinearity of organic EO materials, hence allowing for electro-optic modulation based on the Pockels effect.

SOH phase shifters on a silicon-on-insulator (SOI) platform are realized as silicon slot waveguides which are formed by two silicon rails and the silicon slot is filled with the organic EO material [18]. The EO material is deposited in a dedicated process step after fabrication of the silicon chips, for example by spin coating. A schematic cross-section of an SOH modulator is depicted in Figure 2.10 (a). The buried oxide (BOX) acts as a lower cladding material and the EO material as an upper cladding material. The silicon rails are connected to the electrodes via thin silicon slabs. The slabs are *n*-doped and hence conductive. Dependent on the fabrication process, the structured silicon layer may be covered with a protective silicon oxide cladding layer. In this case, additional vertical vias may be used to connect the silicon slabs and the electrodes and an oxide opening process is required to access the slot for the deposition of the EO material. Detailed explanations on the fabrication processes and SOH modulators in general are found in References [60] and [21]. It shall be noted that while SOH phase shifters rely on slot waveguides, the feeding waveguides can be realized as common silicon strip waveguides that are connected to the slot waveguides via mode converters [69].

For a quasi-TE mode in the slot waveguide, the light is highly confined to the slot region A_{slot} [70]. This is due to the discontinuity of the normal field component at the interface of the silicon rail to the EO cladding material leading to a field enhancement in the slot region. A simulation of the x-component of the optical mode is shown in Figure 2.10 (b). Similarly, a modulating voltage U_{m} applied between the two metal electrodes drops across the narrow slot leading to a strong electric field in the slot, see Figure 2.10 (c) visualizing the x-component of the modulating electric field.

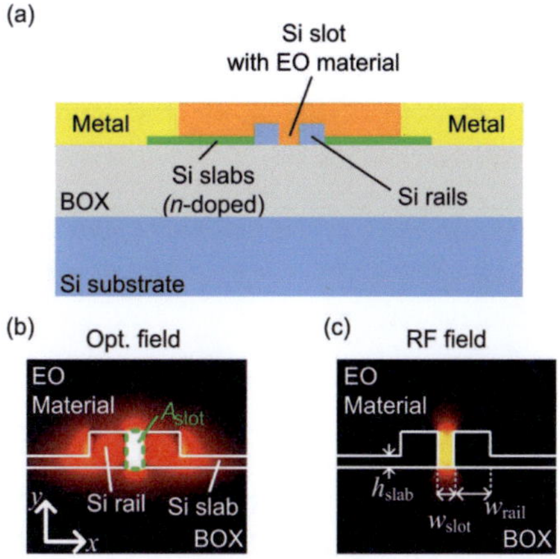

Figure 2.10: Schematic SOH modulators **(a)** cross-section of an SOH phase modulator. The silicon slot waveguide is formed by two silicon rails and filled with the organic electro-optic (EO) material. The rails are connected to metal electrodes via thin slightly *n*-doped silicon slabs. The waveguides structures are realized on a silicon-on-insulator (SOI) wafer with a silicon substrate and a buried oxide (BOX) layer. **(b)** Field distribution of the *x*-component of the optical mode in the slot waveguide. A large portion of the field is confined into the slot area A_{slot} due to the field enhancement at the interface of the silicon rails to the EO material in the slot. **(c)** Field distribution of an electric field induced by a voltage applied to the electrodes. The field is confined to the slot. The large overlap with the optical mode leads to an efficient electro-optic modulation.

The strong overlap of the two fields leads to an efficient EO modulation and can be quantified by the so-called field interaction factor Γ. Using the parallel plate capacitor approximation, the modulating field in the slot region A_{slot} can be assumed to be constant, $E_m = U_m / w_{slot}$, where the slot area A_{slot} is defined by slot width w_{slot} and the height of the silicon rails h_{rail}. The field interaction factor can be defined using the vectorial mode fields $\underline{\mathcal{E}}_0(x,y)$ and $\underline{\mathcal{H}}_0(x,y)$ of the fundamental waveguide modes [17], [21]

$$\Gamma = c\,\varepsilon_0 n_{EO} \frac{\iint\limits_{A_{slot}} \left|\underline{\mathcal{E}}_{0,x}(x,y)\right|^2 \mathrm{d}x\mathrm{d}y}{\iint \mathrm{Re}\left\{\underline{\mathcal{E}}_0(x,y) \times \underline{\mathcal{H}}_0^*(x,y)\right\} \mathrm{e}_z \,\mathrm{d}x\mathrm{d}y}, \tag{2.40}$$

where n_{EO} is the refractive index of the organic cladding material in the slot, c is the vacuum speed of light, and ε_0 is the vacuum dielectric constant. Using the assumption of a constant modulating electric field, Eq. (2.40) is a good approximation, only underestimating the exact value by a few percent [21].

The phase shift in a single SOH phase modulator can be described using the field interaction Γ and the refractive index n_{EO} of the specific EO material and the electro-optic coefficient r_{33},

$$\varphi = \frac{1}{2} n_{\mathrm{EO}}^3 r_{33} E_{\mathrm{m}} \Gamma k_0 L . \tag{2.41}$$

Further, the previous equation uses the phase shifter length L, the modulating field E_{m} and the vacuum wavenumber k_0. The π-voltages for a single SOH phase modulator and for an SOH MZM can then be written as

$$U_{\pi,\mathrm{PM}} = \frac{w_{\mathrm{slot}} \lambda}{\Gamma n_{\mathrm{EO}}^3 r_{33} L} \tag{2.42}$$

and

$$U_{\pi} = U_{\pi,\mathrm{MZM}} = \frac{w_{\mathrm{slot}} \lambda}{2 n_{\mathrm{EO}}^3 r_{33} \Gamma L}, \tag{2.43}$$

where we again drop the subscript 'MZM' for the π-voltage of a MZM in push-pull operation. For the remainder of this thesis, we will refer to SOH MZMs unless noted otherwise. SOH MZMs are realized with electrodes arranged in ground-signal-ground (GSG) configuration, Figure 2.11. SOH MZM typically exhibit π-voltage-length products of the order of 1 Vmm and have been demonstrated down to 0.5 Vmm [20], [71] and 0.32 Vmm [72], respectively.

The second key component of SOH modulators beside the slot waveguide is the organic electro-optic material. These materials exhibit an intrinsic bandwidth in the order of tens of THz and simultaneously provide large electro-optic coefficients up to $r_{33} \approx 500\,\mathrm{pm/V}$ [73]. Due to their good processability and opportunity for deposition on various material systems, organic materials are attractive for the realization of electro-optic modulators. The nonlinearity of the overall material is provided by the electro-optic organic molecules, also referred to as chromophores. These molecules usually consist of a donor and an acceptor part connected by a π-conjugated electron bridge. A detailed explanation on organic nonlinear materials is found in References [73] and [74].

The organic EO material is deposited on the slot waveguides and, subsequently, the chromophores are randomly oriented. Consequently, no appreciable r_{33} is found. To enable a macroscopic nonlinearity, the chromophores require an average acentric order, which can be obtained through a one-time electric field poling process: At an elevated temperature near the glass transition temperature of the EO material, a poling voltage U_{pol} is applied across the floating ground electrodes of an SOH Mach-Zehnder modulator, Figure 2.11. Thereby, poling fields orientated in x-direction with approximately equal strength are induced in the two phase shifter sections, indicated by the green arrows in the figure.

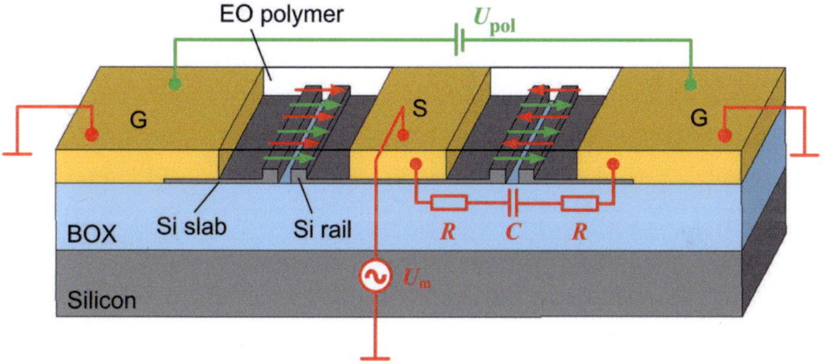

Figure 2.11: Operation principle of SOH MZM. A poling voltage U_{pol} applied across the (floating) ground electrodes generates electric fields that are oriented in the same direction in the two slots (green arrows). The electro-optic (EO) chromophores align accordingly. After poling, when connected to the RF circuit (red) the modulating voltage U_m generates electric fields in the same and opposite direction with respect to the chromophore orientation in the two slots (red arrows). Thereby, phase shifts of equal magnitude but opposite sign are generated and the MZM is operated in push-pull configuration. The dynamic behavior can be understood when interpreting the silicon slabs as resistors and the slot as a capacitance. The *RC* characteristic, microwave losses and a potential velocity mismatch between the optical and electrical mode lead to bandwidth limitations of SOH modulators.

The electric poling fields apply a torque on the dipolar chromophores leading to an orientation of the chromophores according to the poling fields and to an average acentric orientation. While maintaining the poling voltage, the device is then cooled to ambient temperatures. Thereby, the chromophore alignment remains even after removing the poling voltage. During operation, for SOH MZMs with electrodes in GSG configuration, the modulating fields induced by the modulating voltage (or drive voltage) U_m are oriented in the same and opposite direction with respect to the chromophore alignment in each of the two slot waveguides as indicated by the red arrows in Figure 2.11. Thereby, the phase shifts in the two phase modulator sections are of equal magnitude but opposite sign and the SOH MZM is operated in push-pull mode.

SOH modulators have been demonstrated with different EO materials [20], [71], [72], [75], [76]. Properties of various EO materials are discussed in detail in References [20], [77]. For SOH modulators, a record-high in-device EO coefficient of 359 pm/V was reported in [72]. The present work relies on the use of the commercially available organic EO materials SEO100 and SEO250 from Soluxra [78]–[82]. Being guest-host materials comprising amorphous polycarbonate and electro-optic chromophores with a similar chemical structure, both materials exhibit comparable optical properties and glass transition temperatures [78], [82]. While exhibiting a lower electro-optic coefficient r_{33} of 191 pm/V (147 pm/V) in bulk material (in-device) at a wavelength of 1550 nm, their glass transition temperature of approximately 140 °C is more than 50 °C higher compared to the high-r_{33} material used in reference [72]. This is of particular interest concerning the materials' stability: Higher glass transition temperatures are required for operation in environments with elevated temperatures. In terms of the EO materials' stability, one generally discriminates two major effects: stability against thermal relaxation and robustness against photon-induced oxidation.

Thermal relaxation describes the process of decreasing average acentric order in a poled EO material. It is a naturally occurring process since the acentric orientation is energetically unfavorable for the EO chromophores. Thermal relaxation is mainly driven by temperature, which increases the molecular mobility of the chromophores, but it can be accelerated by applying a modulating field. In a previous experiment, it was shown that SOH modulators functionalized with SEO100 are suited for operation at elevated temperatures of 80°C under ambient atmospheric conditions [22] and the manufacturer specifies stability up to 85°C [80]. It is to be expected that even higher operating temperatures can be achieved, for instance, using cross-linking techniques [83] which establish additional chemical bonds subsequent to the poling procedure such that the movement of the chromophores is more strongly impeded. Alternatively, using materials with intrinsically higher glass transition temperatures such as side-chain EO polymers [84] reduces the mobility of the chromophores even at elevated temperatures. Thermal relaxation is a reversible process such that by re-poling the acentric order can be re-established [80].

Secondly, photo-oxidation, which is irreversible, directly affects the molecules themselves and degrades (or destroys) their electro-optic activity. Photon-induced degradation, or photobleaching, of organic EO material, as the name says, is provoked by the absorption of photons and a subsequent chemical reaction with dissolved oxygen. Thereby, it directly affects the chemical structure of the molecule and consequently its optical properties. The chemical relations and a more detailed description of photobleaching is given in Reference [85]. While the EO materials typically do not absorb in the wavelength region around 1550 nm [77], [79], [86], absorption cannot be excluded: two-photon absorption (TPA) can excite higher states and contribute to photobleaching if the intensities are high enough. Due to the nanoscopic cross-section of the SOH slot-waveguide, even moderate power levels of a few milliwatts result in intensities in the order of MW/cm^2 and consequently TPA cannot be excluded.

These aspects are subject to ongoing research activities, and it is expected that photobleaching can be reduced by cross-linking, which reduces oxygen diffusion into the material [87], [88]. Alternatively, encapsulation of the devices using techniques similar to those applied to organic light-emitting diodes (OLED) would eliminate the possibility for oxygen-mediated degradation. Such techniques would make the devices also tolerant towards elevated humidity levels, even though previous investigations did not show humidity-induced material degradations [89].

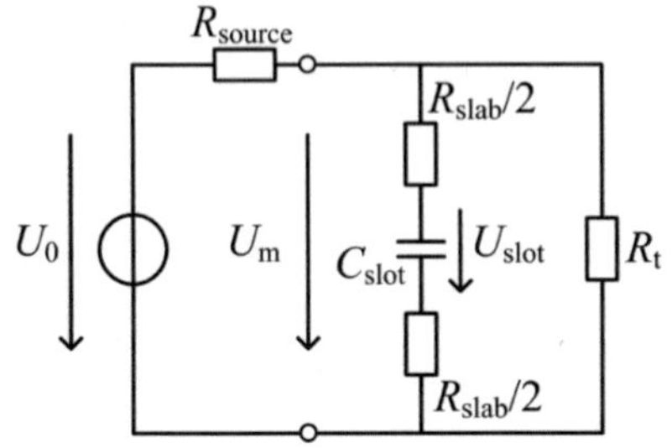

Figure 2.12: Equivalent circuit representation for lumped element approximation of an SOH phase modulator. The open loop voltage U_0 and the source resistance R_{source} represent the voltage source. The slab resistivity R_{slab} for both slabs and the slot capacitance C_{slot} model the modulator which is terminated by a resistance R_t.

2.2.2.1 Bandwidth of SOH modulators and gate voltage

For high-speed communications, it is of great interest to understand the bandwidth limitations of the modulator. For SOH modulators, we can interpret the slot waveguide as a capacitance that is charged and discharged via the (limitedly) conductive silicon slabs as indicated in Figure 2.11. Therefore we find a fundamental *RC* lowpass characteristic resulting from the device principle and the associated geometry. The equivalent circuit representation of an SOH phase modulator in lumped element approximation is depicted in Figure 2.12. Considering the resistivity of the slabs and the terminating resistance, R_{slab} and R_t, the impedance of the slot Z_{slot} itself, and the source voltage U_0, the voltage across the slot U_{slot} in an SOH phase modulator has been derived as [21]

$$\left|U_{\mathrm{slot}}\right| = U_0 \frac{\left|Z_{\mathrm{slot}}(\omega_m)\right|}{\left|2R_{\mathrm{slab}} + 2Z_{\mathrm{slot}}(\omega_m) + R_{\mathrm{t}}\right|}. \qquad (2.44)$$

This derivation assumes that the terminating resistance is matched to the one of the source. Other factors that influence the bandwidth are microwave losses on the transmission line and a potential group velocity mismatch of the electrical and optical wave.

These effects are discussed in more detail in [60] and [21]. According to [21], microwave losses can be included by considering an effective length of the phase modulator,

$$L_{\text{eff}}(\omega_{\text{m}}) = \int_0^L \mathrm{e}^{-\frac{\alpha_{\text{L}}(\omega)}{2}z}\,\mathrm{d}z = \frac{2}{\alpha_{\text{L}}(\omega)}\left(1-\mathrm{e}^{-\frac{\alpha_{\text{L}}(\omega)}{2}L}\right), \tag{2.45}$$

where $\alpha_{\text{L}}(\omega_m)/2$ is the amplitude attenuation factor. For short single-drive SOH MZM we can assume that the group velocity mismatch of the optical and electrical field can be neglected. Then, for a sinusoidal modulating voltage at angular frequency ω_{m} the phase modulation can be expressed as [21]

$$\varphi_{\text{MZM}}(\omega_{\text{m}}) = \varphi_0 \frac{1}{\left|1+\dfrac{R_{\text{slab}}+R_{\text{t}}}{Z_{\text{slot}}(\omega)}\right|} \frac{L_{\text{eff}}(\omega_{\text{m}})}{L}, \tag{2.46}$$

where the phase modulation at modulating angular frequency $\omega_{\text{m}}=0$ is $\varphi_0=\varphi(\omega_{\text{m}}=0)$.

In principle, the device parameters such as the slot capacitance and the resistivity of the slabs and hence the *RC* device bandwidth are defined by design and fabrication. Nevertheless, the device bandwidth can still be modified by applying a gate voltage. For the current designs of SOH modulators, the gate voltage is applied between the device layer and the bulk silicon. Thereby it generates a highly conductive electron accumulation layer below the Si slab surface near the SiO_2-layer which leads to an increased conductivity of the silicon slabs. Consequently, the reduced *RC* constant leads to a higher device bandwidth. The influence of the gate field on the device bandwidth is discussed in Reference [90] and SOH modulators have been demonstrated up to an electro-optic bandwidth of 100 GHz [19]. Similar to the discussion of all-silicon modulators in the previous chapter, the additional free carriers in the silicon slabs lead to an (static) excess loss of the optical waveguides. The effect of the gate voltage and the corresponding electron accumulation layer on excess loss is discussed in detail in References [91]–[93].

As the gate voltage drops across the thin slabs and the bulk oxide layer, typical gate fields in the order of 0.1 V/nm require gate voltages in the order of 100 V. Such high voltages are not a viable option for commercial applications. In Reference [94], this issue is addressed and it proposes a top gate solution: rather than applying the voltage across the oxide layer, the fabricated structure could be complemented with an additional insulating and a silicon gate layer. Thereby, the voltage would not drop across a few micrometers but across a thin layer only.

2.2.3 Plasmonic-organic hybrid (POH) modulators

Besides SOH integration, the related concept of plasmonic-organic hybrid (POH) integration equally relies on a slot waveguide and the interaction of the optical wave with an organic EO material. In contrast to SOH modulators, POH modulators use metallic slot waveguides rather than silicon slot waveguides. In this chapter, the fundamentals of POH modulators are reviewed. A more detailed description of POH modulators can be found in References [95]–[98]. Further, References [17] and [99] review and compare POH and SOH modulators.

Wave propagation in POH waveguides relies on surface electro-magnetic waves, also referred to as surface plasmon polaritons (SPP), which exist at the interface of two materials with opposite sign for the relative permittivity [95], [100]. For telecommunication wavelengths, this holds for metal-dielectric interfaces, and POH modulators are realized by metal-insulator-metal (MIM) waveguides, see Figure 2.13. In POH modulators found in literature [96]–[98], [101], [102], gold (Au) is used as metallic material while an EO material inside the metal slot is used as the insulator. Inside the plasmonic waveguide, the SPP is strongly confined to the slot region [96] and the strong overlap with the modulating field, which is confined to the slot area as well, leads to π-voltage-length products as small as 60 Vµm [97]. Typical device lengths are in the order 10 µm and the device footprint amounts to a few µm^2, only. Due to the small form factor, the device capacity is in the order of a few fF [96], [97]. At the same time, the resistance in the metallic electrodes is negligible and consequently the *RC* low pass characteristic is no practical limitation [96].

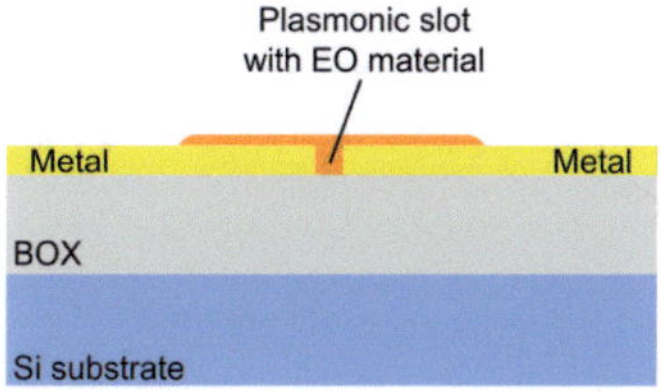

Figure 2.13: Schematic cross-section of a POH phase modulator. The plasmonic slot waveguide is formed by two metal electrodes and is filled with the organic electro-optic (EO) material. Similar to the principle of SOH modulators, the optical mode is highly confined into the slot region allowing for effective EO modulation.

As a result, POH modulators are suited for high-bandwidth applications and device bandwidths of more than 170 GHz have been reported [102].

The carrier-to-sideband power ratio in [102], however, was about 30 dB, indicating that the drive voltage in the order of 0.5 V amounted to only a few percent of the π-voltage. This is a consequence of the comparatively high π-voltages which are typically in the order of multiple volts or exceed 10 V [97], [102]. The considerable propagation losses in the order of 400 dB/mm in plasmonic waveguides with Au electrodes [101] pose limits on the maximum phase shifter length [17] which in turn leads to higher π-voltages than for SOH modulators. Therefore, Reference [17] uses another figure of merit, $aU_\pi L$, which combines the propagation loss a, given in dB/mm, and $U_\pi L$. The figure $aU_\pi L$ is given in dB V and can be interpreted as the π-voltage for a device with a 1 dB insertion loss or as the insertion loss of a device with a π-voltage of 1 V.

Assuming that the propagation loss of a POH device can be reduced to 200 dB/mm when using silver (Ag) electrodes [17] and using a $U_\pi L$-product in the order of 50 Vµm, one finds 10 dB V for the $aU_\pi L$ figure of merit. This compares well to Si *pn*-modulators which are in the same order of magnitude [17]. However, this value is significantly higher than for SOH modulators: When assuming an achievable propagation loss around 2 dB/mm in the Si slot waveguide and taking into account $U_\pi L$-products of 0.5 Vmm [20], [71], we find $aU_\pi L$ to be in the order of 1 dB V [17] for SOH modulators.

3 High-speed data transmission SOH Mach-Zehnder modulators

This chapter discusses the potential of SOH MZM for 100 Gbit/s OOK. This chapter is taken from [J19] which is published in *Scientific Reports*. In order to fit the structure and layout of this document, it was adapted accordingly.

[start of paper [J19]]

Silicon-Organic Hybrid (SOH) Mach-Zehnder Modulators for 100 Gbit/s On-Off Keying

Scientific Reports, Vol. 8, Article no. 2598 (2018)
DOI: 10.1038/s41598-017-19061-8

S. Wolf,[1†] H. Zwickel,[1†] W. Hartmann,[1,2,3†] M. Lauermann,[1]
Y. Kutuvantavida,[1,2] C. Kieninger,[1,2] L. Altenhain,[4] R. Schmid,[4]
J. Luo,[5] A. K.-Y. Jen,[5] S. Randel,[1] W. Freude,[1] and C. Koos[1,2]

[1] Karlsruhe Institute of Technology, Institute of Photonics and Quantum Electronics, 76131 Karlsruhe, Germany
[2] Karlsruhe Institute of Technology, Institute of Microstructure Technology 76344 Eggenstein-Leopoldshafen, Germany
[3] Now with: Physikalisches Institut, University of Muenster, 48149 Muenster, Germany
[4] Micram Microelectronic GmbH, 44801 Bochum, Germany
[5] Department of Chemistry, City University of Hong Kong, Kowloon, Hong Kong SAR
[†] These authors contributed equally to the work

Electro-optic modulators for high-speed on-off keying (OOK) are key components of short- and medium-reach interconnects in data-center networks. Small footprint, cost-efficient large-scale production, small drive voltages and ultra-low power consumption are of paramount importance for such devices. Here we demonstrate that the concept of silicon-organic hybrid (SOH) integration perfectly meets these challenges. The approach combines the unique processing advantages of large-scale silicon photonics with unrivalled electro-optic (EO) coefficients obtained by molecular engineering of organic materials. Our proof-of-concept experiments demonstrate generation and transmission of OOK signals at line rates of up to 100Gbit/s using a 1.1mm-long SOH Mach-Zehnder modulator (MZM) featuring a π-voltage of only 0.9V. The experiment represents the first demonstration of 100 Gbit/s OOK on the silicon photonic platform, featuring the lowest drive voltage and energy consumption ever demonstrated for a semiconductor-based device at this data rate. We support our results by a theoretical analysis showing that the nonlinear transfer characteristic of the MZM can help to overcome bandwidth limitations of the modulator and the electric driver circuitry. We expect that high-speed power-efficient SOH modulators may have transformative impact on short-reach networks, enabling compact transceivers with unprecedented efficiency, thus building the base of future interfaces with Tbit/s data rates.

3.1 Introduction

Global data traffic continues to grow at double-digit annual rates [4], driven by cloud-based service delivery, video on demand, or Internet-of-Things (IoT) applications. To keep pace with this evolution, transceivers on all levels of optical networks are subject to the same challenge: To radically increase data rates, while maintaining acceptable technical complexity and energy consumption. In long-reach core and metropolitan networks, this challenge can be met by advanced modulation formats that exploit advances in high-speed digital signal processing (DSP) and advanced photonic integration. These approaches have led to transmission demonstrations at data rates in excess of 500 Gbit/s on a single polarization, using, e.g., 64-state quadrature amplitude modulation (64QAM) in conjunction with symbol rates up to 100 GBd [45], [103]. When it comes to short-reach transmission over distances of a few kilometers or less in data centers or campus-area networks, however, higher-order modulation formats and the associated coherent reception techniques are prohibitive in terms of technical complexity and power consumption of the DSP. As a consequence, short-reach transmission largely relies on simple on-off-keying (OOK) as a modulation format that can be directly detected with a single high-speed photodiode. Such schemes are at the heart of current transceivers, e.g., for 100 Gbit/s Ethernet interfaces, that usually exploit low-cost vertical-cavity surface emitting lasers (VCSEL) or silicon photonic modulators to generate four spatially or spectrally separated data streams of 25 Gbit/s [104]–[106]. For future interfaces operating at data rates of 400 Gbit/s, 800 Gbit/s or 1.6 Tbit/s, however, parallelization of 25 Gbit/s channels is not a sustainable option [107]. Instead, lane rates of 100 Gbit/s are considered indispensable to maintain further scalability of throughput in campus-area and data-center networks. In this context, the key challenge is to build transmitters that can generate 100 Gbit/s OOK data streams at lowest possible power consumption, and such transceivers have even been identified as the "Holy Grail" of the Ethernet ecosystem by the Ethernet Alliance [107]. In fact, while a wide variety of compact high-speed optical modulators has been demonstrated over the last years [102], [108]–[117], only a few [102], [110], [111], [114] are at all capable of providing OOK data rates of 100 Gbit/s, and often feature comparatively high operating voltages of 2 V or more. When it comes to high-speed Mach-Zehnder modulators (MZM), the most efficient 100 Gbit/s device demonstrated so far is based on 2.5 mm long phase shifters and features a π-voltage of 2 V [111].

In this paper we show that these limitations can be overcome by silicon-organic hybrid (SOH) modulators [17], [19], [20], [22], [71], [72], [75], [76], [94], [118]–[123] that combine the advantages of large-scale silicon photonic integration with the extraordinarily high electro-optic (EO) coefficients obtained by molecular engineering of organic materials [73], [124]. We demonstrate generation of OOK signals up to 100 Gbit/s using a 1.1 mm-long SOH MZM which features a π-voltage of only 0.9 V. The associated switching energy amounts to less than 98 fJ/bit – a record-low value for transmission at 100 Gbit/s OOK using semiconductor-based modulators. In our experiments, we operate the device at a peak-to-peak voltage of 1.4 V, thereby exploiting the nonlinear transfer characteristic of the MZM to mitigate impairments by limited bandwidth of the modulator and of the transceiver circuitry, which leads to an estimated BER down to 6.6×10^{-6}. We support our experiments by a theoretical analysis that takes into account the measured transfer function of the modulator and of the driver circuits. Besides signal generation, we transmit a 100 Gbit/s OOK stream over a dispersion-compensated 10 km standard single-mode fiber (SMF) link. This is the first transmission experiment of a 100 Gbit/s OOK signal generated by a semiconductor-based modulator. Unlike previous implementations [22], [118] and competing device concepts [96], [97], [101], [102] fabricated by high-resolution electron beam lithography, our modulators were processed in a commercial silicon photonics line together with the full portfolio of silicon photonic devices and Ge photodiodes using standard 248 nm deep-UV lithography. In contrast to earlier demonstrations of SOH electro-optic modulators [22], [118], [119], [123], this work focusses on high-speed serial transmission using simple on-off-keying and direct detection techniques rather than technically demanding quadrature amplitude modulation (QAM), thereby considerably reducing hardware complexity and avoiding energy-intensive signal processing. When combined with highly efficient CMOS drivers, SOH modulators have the potential to open a technically and commercially superior avenue towards short-reach transceivers with unprecedented energy efficiency that will be key for future Ethernet interfaces at Tbit/s data rates.

3.2 Materials and methods

3.2.1 Principle of a silicon-organic hybrid (SOH) electro-optic (EO) phase modulator

SOH modulators combine silicon-on-insulator (SOI) slot waveguides and electro-optic (EO) cladding materials [17]. A schematic of an SOH Mach-Zehnder modulator (MZM) and its cross-section are depicted in Figure 3.1 (a) and (b), respectively. The phase shifters consist of a silicon slot waveguide which is formed by two silicon rails [17]. Due to the discontinuity of the normal electric field component at the interface to the silicon rails, the dominant field component $\underline{\mathcal{E}}_{0,x}$ for the quasi-TE polarization in the slot is strongly enhanced, Figure 3.1 (b), Inset (1). The slot is filled with an organic EO material which provides a high $\chi^{(2)}$-nonlinearity (Pockels effect). The silicon rails are connected to aluminum (Al) electrodes via thin n-doped silicon slabs and aluminum vias (not depicted). A modulation voltage applied to the Al electrodes drops completely across the narrow slot and leads to a strong electric field $E_{x,RF}$ which is well confined to the silicon slot region, Figure 3.1 (b), Inset (2). The strong electric field interacts with the EO organic cladding and leads to a pronounced change of the refractive index, and consequently to a phase modulation of the optical wave. For estimating the phase shift φ in an SOH waveguide, we may assume that the modulating radio-frequency RF field features only an x-component which has a constant value of $E_{x,RF} = U_m/w_{slot}$ in the slot region A_{slot}, see Figure 3.1 (b), and which is negligible outside. This leads to the relation [17]

$$\varphi = \frac{1}{2} n_{EO}^3 r_{33} E_{x,RF} \Gamma k_0 L, \tag{3.1}$$

where $k_0 = 2\pi/\lambda$ is the optical wavenumber at a vacuum wavelength λ, Γ is the field interaction factor, n_{EO} is the refractive index of the organic cladding material in the slot if no voltage is applied, r_{33} is the EO coefficient and L describes the length of the phase shifter. The field interaction factor Γ can be calculated in terms of the vectorial mode fields $\underline{\mathcal{E}}_0(x,y)$ and $\underline{\mathcal{H}}_0(x,y)$ of the fundamental waveguide modes [17],

$$\Gamma = \frac{c\varepsilon_0 n_{EO} \iint_{A_{slot}} \left|\underline{\mathcal{E}}_{0,x}(x,y)\right|^2 \mathrm{d}x\mathrm{d}y}{\iint \mathrm{Re}\left\{\underline{\mathcal{E}}_0(x,y) \times \underline{\mathcal{H}}_0^*(x,y)\right\} \mathrm{e}_z \, \mathrm{d}x\mathrm{d}y}, \tag{3.2}$$

Where c is the vacuum speed of light, and ε_0 is the vacuum dielectric constant.

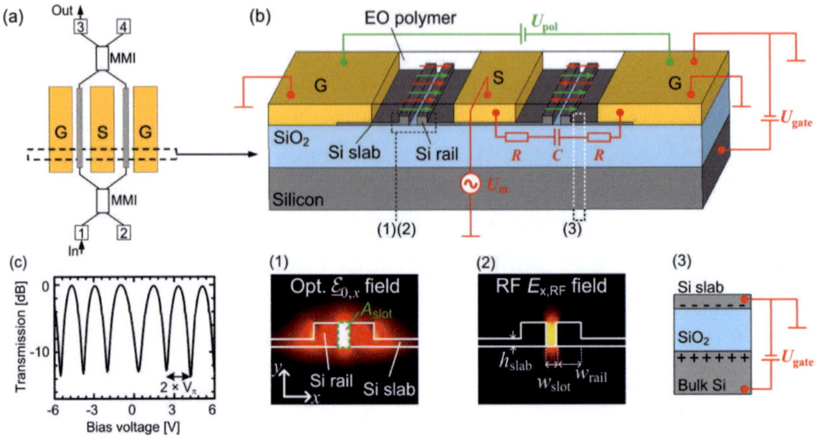

Figure 3.1: Schematic of an SOH Mach-Zehnder modulator (MZM). (a) Top view of an MZM with two SOH phase shifters, a coplanar ground-signal-ground (GSG) transmission line, and a pair of multi-mode interference (MMI) couplers. (b) Cross-sectional view of the phase shifter section of the SOH MZM fabricated on a silicon-on-isolator substrate with a 2 µm thick buried oxide (BOX) layer. The slot waveguide (w_{slot} = 160 nm, w_{rail} = 240 nm, h_{Slab} = 70 nm) is formed by two silicon rails and is embedded into an organic electro-optic (EO) cladding material. The slot is connected to the aluminum (Al) transmission lines by thin *n*-doped silicon slabs and aluminum vias (not depicted). A poling voltage U_{pol} applied across the (floating) ground electrodes at an elevated temperature close to the material's glass transition point aligns the EO chromophores in the slot (green arrows). An electric field generated from a modulation signal U_m applied to the GSG electrodes (red arrows) is oriented parallel (anti-parallel) with respect to the chromophores orientation in the left (right) slots. This leads to a phase shift of equal magnitude but opposite sign in the two slots, resulting in chirp-free push-pull operation. The electronic bandwidth of an SOH MZM is limited by the inherent *RC* lowpass characteristic resulting from the limited conductivity of the *n*-doped silicon slabs and the capacitance of the slot. Inset (1): Dominant *x*-component of the optical electric field $\underline{\mathcal{E}}_{0,x}$ in the slot waveguide with slot area A_{slot}. Inset (2): Field component $E_{x,RF}$ of the electrical RF drive signal. Both the optical and the electrical field are well confined to the slot and overlap strongly for an efficient modulation. Inset (3): Electron accumulation layer. The modulator bandwidth is increased by a decreased resistance (*R*) of the slabs, induced by a charge accumulation layer that can be generated by a "gate voltage" U_{gate} between the bulk silicon and the ground electrodes. (c) DC characteristic of a 1.1 mm long SOH MZM with a π-voltage of 0.9 V measured at a voltage offset slightly more than 2 V. Using this offset avoids screening effects of the applied electric field by free charges in the organic cladding [17].

The dynamic behavior of an SOH device can be understood by considering a lumped-circuit model as indicated in Figure 3.1 (b). It consists of a capacitor C representing the SOH slot waveguide, and of two resistors R which describe the finite conductivity of the n-doped silicon slabs. The resulting RC low-pass characteristic as well as the RF propagation loss lead to bandwidth limitations [19]. The conductivity of the silicon slabs and hence the device bandwidth can be increased by, e.g., applying a voltage between the device layer and the bulk silicon, which leads to an electron accumulation layer and hence to a reduced resistivity of the slabs [92],[19], see Figure 3.1 (b), Inset (3). Using this approach, bandwidths in excess of 100 GHz have been demonstrated [19]. The necessary gate voltage can be diminished by using thin oxide and doped polysilicon gates deposited on top of the slab regions [94].

Note that the SOH approach maintains the full advantages of silicon photonics, exploiting highly mature CMOS processes for fabrication of the slot-waveguide base structures, onto which EO materials are deposited in a highly scalable post-processing step. In particular, SOH devices can be seamlessly integrated into complex photonic integrated circuits (PIC) that exploit the full range of devices available on the silicon photonic platform.

3.2.2 Design, fabrication, and operation of SOH Mach-Zehnder modulators

In this work we use a 1.1 mm long SOH Mach-Zehnder modulator (MZM) formed by two SOH phase shifter sections. The phase shifters comprise 240 nm wide rails and 160 nm wide slots and are covered by the organic EO material SEO100 [80], which has a refractive index of $n_{\text{EO}} = 1.73$. This leads to a field interaction factor $\Gamma \approx 0.16$. Our experiments build upon a series of technological advances in comparison to earlier publications [20], [22], [119]. These advances relate, e.g., to fabrication techniques, to the materials, for which stability has been improved greatly [22], and to the underlying poling procedures. In particular, the silicon photonic base structures in this work are fabricated in a 248 nm deep ultra-violet (DUV) optical lithography process at A*Star IME in Singapore – unlike previous generations of SOH modulators, for which electron-beam lithography was used [22], [118]. Fabrication on a standard platform allows co-integration with the full portfolio of silicon photonic devices. In order to comply with the design rules of the foundry, some adaptations of the modulator and the transmission line were required. As an example, the slots had to be defined on a dedicated mask layer to enable better control of the slot width. Regarding electrical wiring, our current chips comprise two metal layers, rather than a single layer as used in previous devices. The additional wiring layer is embedded into a 2 µm-thick layer of oxide (not depicted in Figure 3.1 (b)). In addition, our current devices rely on a dedicated process to locally remove the top oxide layer in the slot regions, such that the EO material can be deposited onto the pre-fabricated chips in a separate process. The deposition of the EO material is completely independent from the processing in the fab. Note that, besides opening of the top oxide, there is no need to further adapt the fabrication workflow of the silicon photonic base structure. The approach should also be compatible with on-chip integration of driver electronics, and we do not expect any fundamental technology-related roadblocks when it comes to mass production of SOH structures in silicon foundries.

The EO material has a high EO coefficient of 166 pm/V as measured in bulk material for a wavelength of 1550 nm [79]. After spin coating, the macroscopic EO activity of the cladding is activated by poling [71] at an elevated temperature close to the material's glass transition temperature. To this end, a DC poling voltage is applied across the (floating) ground electrodes to align the EO chromophores in the two slots. The direction of alignment is defined by the direction of the electric DC poling field and is the same direction in both slots, indicated by green arrows in Figure 3.1 (b). The poling voltage remains applied while cooling the device to room temperature in order to freeze the chromophores in their state of orientation. Applying a modulation voltage to the signal electrode after poling leads to modulating fields oriented in opposite directions with respect to the chromophore alignment in the two slots. The modulating field is indicated by red arrows in Figure 3.1 (b). This leads to phase shifts of equal magnitudes but opposite signs in the two slots. This results in an efficient push-pull operation [17] and leads to chirp-free amplitude modulation provided that the device is perfectly balanced; see Section D.1 for a more detailed discussion of chirp properties and of the impact of imbalance.

In general, SOH EO modulators stand out due to their high modulation efficiency, which can be expressed by the π-voltage-length product $U_\pi L$, where U_π is the voltage required to achieve a phase difference of π in the two arms of the MZM, and where L denotes the phase shifter length. SOH MZM have been demonstrated with $U_\pi L$ products down to 0.5 Vmm [20], [71] – more than an order of magnitude below that of conventional *pn*-depletion type devices [67], [125]. Using SOH modulators, we have demonstrated optical signal generation with drive voltages down to 80 mV_{pp} and energy consumptions of the order of 1 fJ/bit using OOK modulation [20]. In contrast to plasmonic-organic hybrid (POH) electro-optic modulators [96], [97], [101], which adapt the concept to plasmonic waveguides, SOH devices stand out due to significantly lower propagation losses, which enable larger lengths of phase shifters, and hence lower drive voltages [17]. As a quantitative measure, the product of the π-voltage U_π and the achievable insertion loss aL can be used, where a denotes the propagation loss in the phase shifter in dB/mm and where L is the phase shifter length. For POH modulators, this figure is usually above 10 dB V, whereas values of 1 dB V can be achieved by SOH devices [17]. The capabilities of the SOH platform can further be extended to highly efficient phase shifters based on liquid crystals [120], or to hybrid lasers that exploit light-emitting cladding materials [126].

While the present work concentrates on SOH MZM for high-speed OOK, the SOH modulator concept has also been proven to be perfectly suited for generation of advanced modulation formats such as 4-state pulse-amplitude modulation (4PAM, 120 Gbit/s) [122], 8-state amplitude shift keying [75] (8ASK, 84 Gbit/s), quadrature phase shift keying (QPSK), or 16-state quadrature amplitude modulation [76], [118], [119], [123] (16QAM). For 16QAM signaling, we have recently demonstrated line rates of up to 400 Gbit/s [118], [123]. At the same time, the efficiency of SOH modulators is accentuated by the ability for operation without external drive amplifiers, even for generation of higher-order modulation formats [76], [119], where an electrical energy consumption of down to 18 fJ/bit has been demonstrated for 16QAM signaling. While these modulation formats lead to larger spectral efficiency and consequently to higher data rates, they considerably increase the complexity of transmitter and receiver and are hence not well suited for short-reach transmission in data centers or campus-area networks. This is particularly true for coherent communications where the receiver requires a dedicated photonic integrated circuit with two balanced detectors per polarization, a local oscillator laser, and extended DSP rather than just a simple photodiode.

The MZM used in our experiment features a π-voltage of 0.9 V at a wavelength of 1550 nm, see Figure 3.1 (c), which, for a device phase shifter length of 1.1 mm, corresponds to a $U_\pi L$-product of 1 Vmm. Note that the π-voltage was measured at DC bias voltages above 2 V. For smaller bias voltages, we observe slightly increased spacings of the transmission dips and hence slightly increased π-voltages, which is attributed to free ions in the cladding that lead to a partial screening of the applied fields at small bias voltages [17]. However, this effect is only observable for low frequencies and does not impede RF operation [71].

The slight increase of $U_\pi L$ in comparison to the value reported in references [20], [71] is caused by a reduced r_{33}-coefficient of the presently used EO material, which was selected for high thermal stability rather than for highest EO activity. The EO coefficient r_{33} can be estimated from the measured $U_\pi L$-product of the MZM operated in push-pull, the calculated field interaction factor Γ and from the slot width w_{Slot},

$$U_\pi L = \frac{w_{\text{slot}}\lambda}{2 n_{\text{EO}}^3 r_{33} \Gamma}, \qquad r_{33} = \frac{w_{\text{slot}}\lambda}{2 U_\pi L n_{\text{EO}}^3 \Gamma}. \tag{3.3}$$

In this relation, λ is the carrier wavelength, and n_{EO} is the refractive index of the organic cladding material [17]. The factor of 2 in the denominator results from the push-pull operation. For our device, we find an r_{33} coefficient of 147 pm/V, which compares well to values of 166 pm/V reported for bulk SEO100, see reference [79]. Note that SEO100 was chosen due to its temperature stability [80]. Using specially optimized material systems, we have previously demonstrated even higher EO coefficients in excess of 230 pm/V [20], [72], but these materials did not yet feature thermal stability for extended operation at 85 °C. The modulator structure as well as the organic material are well suited to operate over a large range of infrared telecommunication wavelengths comprising all relevant transmission bands between 1260 nm and 1675 nm, see Section D.2 for details.

For the special SOH devices used in our experiments, rather high optical losses were observed with fiber-to-fiber attenuations of 20 dB or more. These high losses were caused by a fabrication problem, which led to contamination of the slot waveguides with Germanium residuals – this problem was fixed in newer device generations by adapting the process flow. In the presented experiments, the losses of the devices comprise approximately 4.5 dB of fiber-chip coupling loss for each of the grating coupler interfaces (9 dB in total), around 1 dB of excess loss for a pair of strip-to-slot converters, approximately 1 dB of excess loss for a pair of multi-mode interference couplers, and an additional approximately 1 dB for on-chip waveguides. This leaves approximately 8 dB for the 1.1 mm-long slot-waveguide section, corresponding to rather high propagation losses of 7.3 dB/mm for this specific device generation. These losses lead to $aU_{\pi}L$ products of 8 dB V, which is clearly above the 2.8 dB V that were previously demonstrated for SOH devices [20], but still well below the approximately 25 dB V found for POH modulators [77].

Another important performance parameter of MZM is the static extinction ratio $\delta^{(\text{stat})}$, which is defined by the squared sum of the superimposed optical field strengths of both arms at the output of the MZM, related to the squared difference of the field strengths (or amplitudes),

$$\delta^{(\text{stat})} = \frac{\left(A_1 + A_2\right)^2}{\left(A_1 - A_2\right)^2}. \tag{3.4}$$

If the splitting and combining ratios are not exactly 50/50 or if the loss in one arm is different from the loss in the other arm, then the amplitudes of the superimposed fields are different, and the extinction ratio (ER) is finite. This may lead to a chirped output signal, even if the modulator is operated in push-pull mode, see Section 3.3 and Section D.1 for details. The (static) ER can be determined by measuring the transmission of the MZM as a function of the applied DC voltage, see Figure 3.1 (c). For the device used in the transmission experiments, the ER is rather low and amounts to approximately 14 dB. We attribute this to deviations of the MMI coupler from the ideal splitting ratio and to unequal propagation losses in the two arms of the MZM caused by the Germanium contaminations of the slot waveguides. Note that, due to the specific chip design, ports 2 and 4 of the MZM were inaccessible, see Figure 3.1 (a). Using ports 1 and 4 or ports 2 and 3 instead would eliminate the problem of non-ideal MMI couplers. In general, SOH devices can provide ER which are much better than the 14 dB obtained here – for other devices with similar device layouts we typically measure ER of $\delta^{(\mathrm{stat})}$=(20…30)dB, see Section D.1 and reference [71]. Note that the "dynamic" extinction ratio $\delta^{(\mathrm{dyn})}$ of the data signal is not only dictated by the static extinction ratio $\delta^{(\mathrm{stat})}$ of the modulator itself, but also by other effects such as inter-symbol interference, see Section 3.3.2 below.

3.3 Results and discussion

3.3.1 Setup for signal generation

The experimental setup for data signal generation is depicted in Figure 3.2 (a). An external cavity laser (ECL) provides the optical carrier at a wavelength of approximately 1550 nm. The light is coupled to and from the SOH MZM via grating couplers. While these couplers are perfectly suited for testing, they introduce limitations of the operating wavelength range. In advanced device implementations, these grating couplers might be replaced by edge coupling, e.g., based on free-space assemblies of micro-lenses and prisms [127], or with 3D free-form waveguides or lenses printed by two-photon polymerization [128]–[132]. After modulation, the optical signal is fed into an erbium-doped fiber amplifier (EDFA) with a gain of 35 dB, an optional 10 km long fiber, and a 2 nm wide optical band-pass filter to remove out-of-band noise, before being detected by a 100 Gbit/s photodiode. At the transmitter, an arbitrary waveform generator (AWG, Keysight M8195A) is used to synthesize the electrical drive signals, using two independent pseudo-random binary sequences (PRBS) of length 2^9-1. These signals enter a 2:1 electrical multiplexer (MUX, SHF 603A), the output of which is a non-return-to-zero (NRZ) signal with a peak-to-peak voltage swing of 0.4 V_{pp} and double the symbol rate of the inputs. A radio-frequency (RF) amplifier (SHF 827) with a nominal bandwidth of 70 GHz is used to boost the signal at the MUX output to a peak-to-peak voltage of 1.4 V_{pp}. This signal is then coupled to the ground-signal-ground (GSG) transmission line of the MZM via microwave probes having a nominal bandwidth of 67 GHz. A DC bias voltage is applied via the same microwave probe to set the operating point of the MZM to the quadrature (3 dB) point. An external 50 Ω termination resistor connected to the end of the transmission line via a second microwave probe prevents back-reflections. Because de-multiplexers operating at 100 Gbit/s are commercially not yet available, we could not analyze the data stream in real-time, and it was hence impossible to measure the bit error ratio (BER) directly. Instead, the received electrical signal is analyzed using an Agilent 86100C digital communications analyzer (DCA) with a 70 GHz equivalent-time sampling module (Agilent 86118A). We record the electrical eye diagrams and extract the quality factor (Q-factor) which is defined by the signal's mean levels u_1 and u_0 for the logical '1' and the logical '0', and by the corresponding standard deviations σ_1 and σ_0,

$$Q = \frac{u_1 - u_0}{\sigma_1 + \sigma_0}. \tag{3.5}$$

The BER of the data signal can be estimated from the measured Q-factor [36]

$$\mathrm{BER_e} = \frac{1}{2}\mathrm{erfc}\left(\frac{Q}{\sqrt{2}}\right). \tag{3.6}$$

The complementary error function is defined by

$$\mathrm{erfc}(z) = \left(2/\sqrt{\pi}\right)\int_z^{\infty} \exp\left(-t^2\right)\mathrm{d}t .$$

3.3.2 Generation and transmission of 100 Gbit/s OOK

The NRZ eye diagrams of the OOK drive signals for different data rates are depicted in Figure 3.2 (b), (c). The clearly open eyes of the MUX output are shown in the first column, Figure 3.2 (b), while the eye diagrams of the RF drive amplifier output are depicted in the second column, Figure 3.2 (c).

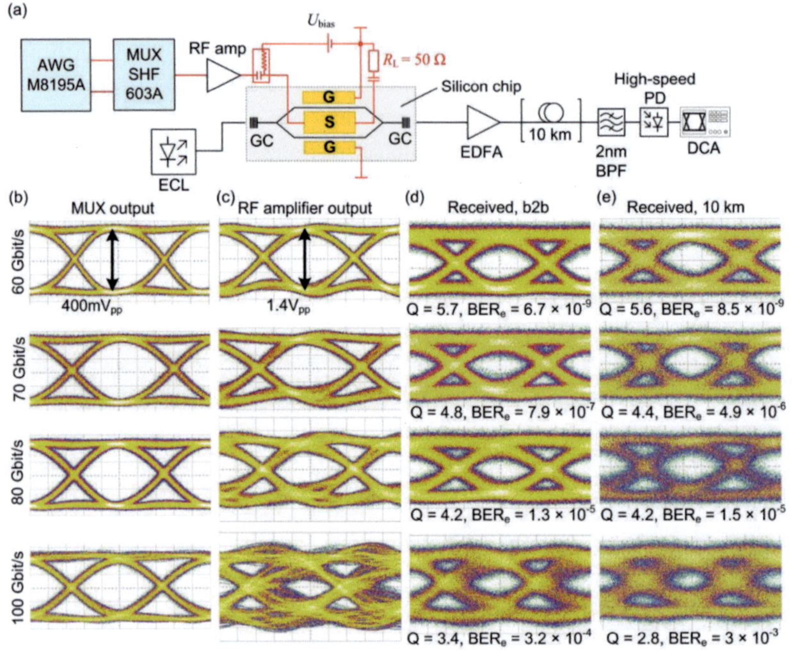

Figure 3.2: Experimental setup and measured eye diagrams. (a) Setup for 100 Gbit/s OOK data generation. An external cavity laser (ECL) provides the optical carrier. Optical power is coupled to and off the 1.1 mm long SOH MZM chip via grating couplers (GC). An erbium-doped fiber amplifier (EDFA) with a 2 nm optical bandpass filter (BPF) compensates chip losses. An optional dispersion-compensated 10 km fiber link is used for transmission experiments. A 100 Gbit/s photodiode detects the signal and feeds it to a 70 GHz equivalent-time sampling module of a digital communications analyzer (DCA). The electrical drive signal is derived from two independent pseudo-random binary sequences (PRBS) generated in an arbitrary waveform generator (AWG, Keysight M8195A). These sequences are fed to a multiplexer (MUX, SHF 603A), combined to a single binary NRZ sequence (≤ 100 Gbit/s), amplified with a radio-frequency (RF) amplifier, and coupled to the chip using microwave probes. A bias-T adds a DC voltage U_{bias} to set the MZM to the quadrature (3 dB) operating point. The ground-signal-ground (GSG) transmission line is terminated with a 50 Ω resistor. (b) Eye diagrams of MUX output (voltage swing 400 mV$_{pp}$) for data rates 60 Gbit/s, 70 Gbit/s, 80 Gbit/s, and 100 Gbit/s. (c) Eye diagrams of RF amplifier output (voltage swing 1.4 V$_{pp}$). The amplifier transfer function causes signal distortions. (d) Eye diagrams, measured Q-factor and estimated BER$_e$ after detection (back-to-back, b2b) and (e) after transmission over the dispersion-compensated 10 km long link. In the b2b measurement, a gate field was applied for the 100 Gbit/s signal only. For the transmission, a gate field was applied for both the 80 Gbit/s and 100 Gbit/s signal.

Up to data rates of 60 Gbit/s, the eyes are well open, while at 80 Gbit/s the signal starts deteriorating. This is due to the bandwidth limitations of the amplifier, which predominantly affect its phase response and lead to significant group delay dispersion, see Section 3.3.3 for details.

In a back-to-back measurement without the 10 km long transmission fiber, we record the eye diagrams of the optical signals after direct detection of data signals from 60 Gbit/s to 100 Gbit/s, Figure 3.2 (d). For data rates of 60 Gbit/s and 70 Gbit/s, we measure open eyes and Q-factors of Q = 5.7 and 4.8, respectively. Following Eq. (3.6), we estimate BER_e of 6.7×10^{-9} and 7.9×10^{-7}. At 80 Gbit/s and 100 Gbit/s, the measured Q-factors of 4.2 and 3.4 correspond to an estimated BER_e of 1.3×10^{-5} and 3.2×10^{-4}, respectively. All of these BER_e values are below the threshold of 4.5×10^{-3} for hard-decision forward error correction (FEC) with 7% overhead [133]. A gate field was applied for the 100 Gbit/s signal only. The 100 Gbit/s achieved in our experiment corresponds to the highest OOK data rate generated by a silicon-based modulator so far, see Section 3.3.6 for a more detailed comparison to other experiments. Note that in our as well as in competing high-speed OOK demonstrations [102], [110], [111], the data rate of 100 Gbit/s does not refer to the net data rate, but to the line rate and hence includes the 7 % FEC overhead.

The rather low static extinction ratio (ER) of approximately 14 dB of our MZM leads to a residual chirp of the generated data signal, which would not occur for perfectly balanced devices. For quantifying the chirp of the data signal, we use the chirp parameter α that is essentially defined by the ratio of the phase modulation to the amplitude modulation [134], [135],

$$\alpha = 2P\frac{\mathrm{d}\varphi/\mathrm{d}t}{\mathrm{d}P/\mathrm{d}t}. \tag{3.7}$$

In this relation, φ denotes the phase and P the time-dependent power of the optical signal averaged over a few optical cycles. In SOH devices, imbalance of the MZM arms is the dominant source of chirp, see Section D.1 for a more detailed discussion and an experimental verification. The magnitude of the chirp parameter α can thus be directly related to the ratio $\gamma = A_2/A_1$ of the fields in the two MZM arms [134]

$$|\alpha| = \frac{1}{\gamma} \frac{\varphi_1 + \gamma^2 \varphi_2}{\varphi_1 - \varphi_2}. \tag{3.8}$$

In this relation, φ_1 and φ_2 denote the phases shifts in the individual MZM – for push-pull modulation we can assume $\varphi_2 = -\varphi_1$. The field amplitude ratio γ can be derived from a measurement of the (static) extinction ratio $\delta^{(\text{stat})}$ defined in Eq. (3.4),

$$\gamma = \frac{\sqrt{\delta^{(\text{stat})}} - 1}{\sqrt{\delta^{(\text{stat})}} + 1}. \tag{3.9}$$

Note that the electric fields are chosen such that $A_1 > A_2$ and hence $0 \leq \gamma \leq 1$. As a consequence, Eq. (3.8) only allows determining the magnitude of α, but not its sign – this would require knowing whether the stronger optical amplitude is associated with the MZM arm having a positive or a negative phase shift, which cannot be derived from a measurement of the static extinction ratio. Using Eqs. (3.8) and (3.9), the static extinction ratio of 14 dB of our device translates into a magnitude of the chirp parameter of $|\alpha| \approx 0.42$, which is well below chirp parameters of $|\alpha| = 0.8$ that are obtained for conventional *pn*-depletion type silicon modulators with comparable ER [136].

We also measure the dynamic extinction ratios of our data signals. For 100 Gbit/s signaling, the measured dynamic ER amounts to 5 … 7 dB, which compares well to a measured extinction ratio of 6.1 dB that was observed for a conventional *pn*-depletion type silicon modulator [115] at 70 Gbit/s. Note that the dynamic extinction ratio measured from the eye diagram of a data signal is generally worse than the static extinction ratio of the underlying modulator according to Eq. (3.4). This is due to inter-symbol interference and quadratically detected optical noise.

In addition to the back-to-back experiment, we transmit the data signals over a dispersion-compensated fiber link of 10 km standard single-mode fiber (SMF) having a negligible residual dispersion of 2.6 ps/nm. The eye diagrams of the data signals received after transmission are depicted in Figure 3.2 (e). For the 60 Gbit/s and 70 Gbit/s data, the Q-factors of 5.6 and 4.4 did not significantly deteriorate compared to the Q-factors of 5.7 and 4.8 in the back-to-back measurements. These Q-factors correspond to $BER_e = 7.9 \times 10^{-7}$ and $BER_e = 4.9 \times 10^{-6}$, respectively. For data rates of 80 Gbit/s and 100 Gbit/s, a gate field was applied and the measured Q-factors of 4.2 and 2.8 correspond to BER_e values of 1.5×10^{-5} and 3.0×10^{-3}. This experiment corresponds to the first 100 Gbit/s OOK transmission demonstration using a modulator on the silicon photonic platform. The transmission demonstration over the dispersion-compensated 10 km fiber link was performed with binary drive signals without further signal processing. Still, the results compare very well to recently published demonstrations of 100 Gbit/s OOK transmission over an uncompensated 1.8 km-link using InP based devices in combination with digital equalization at the receiver [111].

3.3.3 Theoretical analysis of bandwidth limitations

To analyze the impact of bandwidth limitations on our experiments, we reproduce the results by simulations, see Figure 3.3. To this end, we use a vector network analyzer (VNA) to measure the frequency response of a 1.1 mm long SOH MZM with a gate field of 0.1 V/nm, as used for the 100 Gbit/s transmission experiment. The VNA generates an input signal with varying frequency, which is coupled to the (terminated) modulator. The modulated optical power is received with a photodiode with a calibrated frequency response, which feeds its output back to the VNA for a characterization of the electro-optic-electric (EOE) bandwidth. The measured frequency response (by modulus and phase) is depicted in Figure 3.3 (c). The 6 dB point [137] of the EOE frequency response is found to be about 25 GHz.

At first sight, a 6 dB EOE bandwidth of 25 GHz seems rather small for generating a 100 Gbit/s NRZ signal. In general, considering a low-pass filter at the receiver, an optimum signal-to-noise power ratio is found for a 3 dB bandwidth of approximately 65 % of the symbol rate [138]. This bandwidth leads to an ideal trade-off of noise power in the filter passband and inter-symbol interference (ISI) caused by the limited filter bandwidth. In our case, however, the situation is different: The bandwidth limitation is caused by a low-pass characteristic associated with the electrical part of the modulator, which is then followed by the nonlinear, cosine-shaped MZM intensity transmission characteristic, see Figure 3.3 (d). As a consequence, while low amplitudes of the drive signal translate linearly to optical output power, higher amplitudes are compressed, thereby mitigating ISI-related amplitude fluctuations of the drive signal. To illustrate this effect, we analyze our transmission system in MATLAB. To emulate the limited rise and fall times of the hardware in the simulation, we use cosine-shaped pulses in the time-domain (not to be confused with raised-cosine pulse shaping with a raised-cosine shaped spectrum) to approximate the measured output signal of the MUX, see Figure 3.2 (b). Details can be found in Appendix E.

In the simulation, this signal is then fed to the drive amplifier, which has a gain of 11 dB and which is modeled by the measured S_{21}-parameter, Figure 3.3 (b). At the output of the amplifier, the drive signal features a peak-to-peak voltage swing of 1.4 V_{pp}, which is used as an input signal to the modulator. The modulator is modeled by the measured frequency response depicted in Figure 3.3 (c), followed by the cosine-shape time-domain power transfer function, for which we assume a π-voltage of 0.9 V. The frequency response was set to nearly zero (-200 dB) beyond the measured frequency range. The results of the simulation are depicted in the lower part of Figure 3.3 (a). Interestingly, the electrical eye diagrams at the amplifier output show already significant distortions, see left column of eye diagrams in Figure 3.3 (a), even though the magnitude of the frequency response of the device remains flat up to more than 70 GHz. It turns out that these distortions are not caused by the low-pass characteristic of the amplifier, but by its phase response, which starts dropping by approximately -200 ° between 40 GHz and 75 GHz, see Figure 3.3 (b).

This can be confirmed by assuming a flat phase characteristic in the simulation, for which the distortions disappear. The low-pass modulator frequency response leads to further closure of the eye due to ISI, see center column of eye diagrams in Figure 3.3 (a), and it is only the compression of large amplitudes in the MZM that reproduces an open eye, see right column of eye diagrams in Figure 3.3 (a).

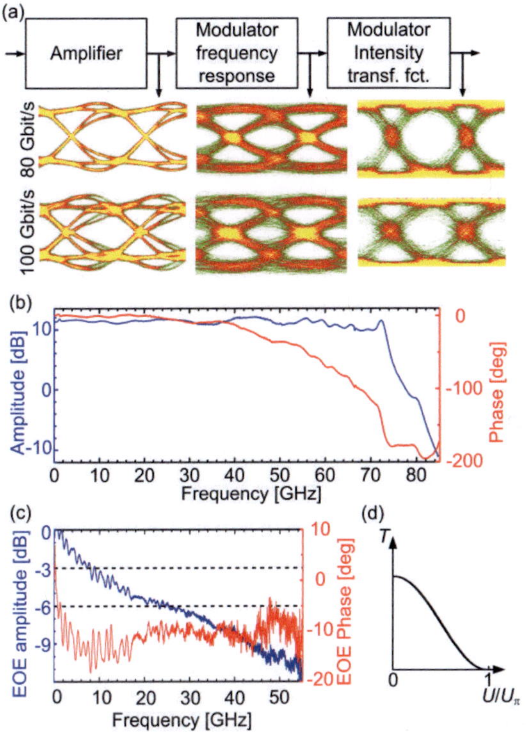

Figure 3.3: Analysis of bandwidth limitations. (a) Simulated eye diagrams. Amplifier input: Cosine-shaped pulses with full-width-half-maximum (FWHM) duration equal to the symbol period. The amplifier is modeled by its measured frequency response, see (b). The modulator is represented by its measured small-signal transfer function, see (c), followed by a cosine-shaped intensity transfer characteristic as sketched in (d). Color map differs from Figure 3.2. (b) Modulus and phase of measured S_{21} parameters of 70 GHz drive amplifier. (c) Modulus and phase of measured small-signal electro-optic frequency response of a 1.1 mm long SOH MZM. Note that the bandwidth of an EO modulator is usually specified by the modulation frequency which corresponds to a 6 dB drop of EOE response [110], [137]. For the spectral component associated with the modulation frequency, this corresponds to a drop of the optical power and hence of the photocurrent amplitude by factor of two, which is measured as a four-fold (6 dB) decrease of the spectral power density by the VNA. For our device, the 6 dB bandwidth amounts 25 GHz. Organic cladding: SEO100. Gate field: 0.1 V/nm. (d) Intensity transfer function T of a MZM vs. normalized drive voltage U/U_{π}.

Using this property allows the generation of high-speed NRZ signals with data rates well beyond the modulator's small-signal 6 dB bandwidth. Note that, in contrast to other 100 Gbit/s OOK demonstrations [102], [111] or high symbol-rate coherent modulation experiments [123], our demonstration does not require any digital pre-distortion or post-equalization and can rely on simple binary drive signals for the modulator at the transmitter and on a simple sampling oscilloscope for measuring the eye diagram at the receiver.

The compression of large amplitudes by the cosine-shaped transfer function is a general property of all MZM. The effect, however, can only be exploited if the available peak-to-peak drive voltage reaches the π-voltage of the device. It is a unique feature of our SOH MZM that this amplitude compression can be achieved at comparatively low drive voltages, which can be realistically generated by currently available driver circuits. The low operating voltage becomes particularly crucial at high data rates, where drive signals with high modulation amplitude are particularly difficult to generate. As an example, conventional *pn*-depletion-type modulators require peak-to-peak drive voltage swings in excess of 5 V to reach compression of the power transfer function, which would be hard to achieve with currently available electrical drivers for 100 Gbit/s, in particular when lower power consumption shall be maintained.

In our experiments, the length of the PRBS was limited to 2^9-1 due to the memory size and the memory granularity of the AWG, but does not represent a fundamental limit of the device. Using a longer PRBS sequence would lead to elongated sequences of subsequent logical '0' or '1' and hence to more low-frequency components in the signal spectrum. This should not affect the results – the lower cut-off frequency of our transmitter amounts to only 70 kHz, dictated by the AC-coupled drive amplifier. To confirm this expectation, we simulated PRBS sequences of lengths 2^9-1 and $2^{18}-1$ – the maximum possible length that could be handled by our computer. We do not find any relevant change of the received eye diagram.

3.3.4 Signal generation using a 100 GSa/s digital-to-analog converter (DAC) and a real-time oscilloscope

In our previous experiments we were unable to measure the BER directly, since the eye diagram at the receiver was taken with equivalent-time sampling technique. For a more direct analysis of the BER, we improved our setup as follows: AWG and MUX were replaced by a programmable DAC with a sampling rate of 100 GSa/s (Micram DAC4) and analog bandwidth above 40 GHz, and the equivalent-time sampling oscilloscope was substituted by a real-time oscilloscope having a bandwidth of 63 GHz. For detection, we use a photodiode with a bandwidth of 70 GHz, similar to the one of our previous experiment. We record the time traces of the photodiode current and analyze it offline with MATLAB. We operate the MZM without a gate field. For a 100 Gbit/s OOK signal we measure a Q-factor of 2.4 and estimate a $BER_e = 8.2 \times 10^{-3}$, Figure 3.4 (a). The Q-factor in the previous experiment was larger and the estimated BER_e was smaller because the MUX acted as a limiter, and because the gate field increased the MZM bandwidth. When taking the directly measured BER from the DAC-based experiment, however, we find a value of only $BER_m = 4.2 \times 10^{-3}$, which is smaller than the BER_e estimated from the measured Q-factor and falls just within the limits of hard-decision (HD) FEC with 7% overhead. This illustrates the issues involved in estimating a BER from the Q-factor using Eq. (3.6).

Note that FEC is presently not used for short-reach optical interconnects because of the computational latency. However, the OpenOptics multi-source agreement (MSA) [139] already describes the use of optimized, low-latency FEC codes that allow to trade coding gain for latency [140] and therefore exhibit more stringent BER thresholds. With a post-equalization filter, we measure a BER_m of 6.6×10^{-6}, see Figure 3.4 (c), which is within the limits of low-latency FEC. On-chip equalizers are commonly used in state-of-the-art transmitter drive chips [141].

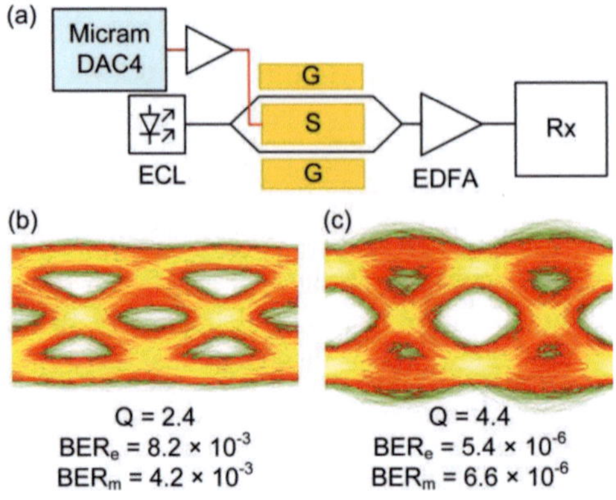

Figure 3.4: Experimental b2b setup and measured eye diagrams for a 100 Gbit/s OOK. (a) A programmable digital-to-analog converter (DAC, 100 GSa/s) serves as an AWG and drives the MZM (no gate field). The receiver Rx comprises a BPF, a photodiode, and a 63 GHz real-time oscilloscope. (b) Received eye diagram, measured Q-factor, estimated BER_e, and measured BER_m at 100 Gbit/s without post-equalization and (c) with post-equalization. Color map differs from Figure 3.2.

3.3.5 Energy considerations

In data center networks, power dissipation is a severe aspect, and the driver electronics of EO modulators play an important role in the overall energy consumption of the transceiver. The design and therefore the power dissipation of the driver depends strongly on the specifications of the modulator. In the following, we analyze the modulator's energy consumption per bit, which we regard as a figure of merit for the overall power dissipation of the transceiver. To this end, we assume that the GSG transmission line impedance is matched to the 50 Ω drive circuitry and to the terminating resistor $R_t = 50\ \Omega$, see Figure 3.2 (a). The modulator is driven with rectangular non-return-to-zero (NRZ) pulse sequences, for which logical '1' and '0' are equiprobable. The energy consumption per bit can then be calculated [20] by dividing the electrical power associated with the drive voltage amplitude by the line rate r,

$$W_{\text{bit}} = \left(\frac{U_{\text{m}}}{2}\right)^2 \frac{1}{R_t r}. \tag{3.10}$$

For $R_t = 50\ \Omega$, a peak-to-peak voltage swing of $U_m = 1.4$ V, and a line rate of 100 Gbit/s, we find an energy consumption of 98 fJ/bit. This is, to the best of our knowledge, the smallest power consumption ever reported for a semiconductor-based MZM at 100 Gbit/s OOK. Note that the biasing and the gate voltage do not involve any measureable DC current flow and hence do not contribute to the energy consumption.

3.3.6 Competitive benchmarking and application potential

The experiments presented in the previous sections demonstrate the unprecedented performance of SOH EO modulators to realize high-speed energy efficient OOK transceivers for data-center and campus area networks. A more detailed comparison to competing concepts of semiconductor-based devices is given in Figure 3.5 (a). The graph shows the required peak-to-peak drive voltages as a measure of the energy consumption in dependence of experimentally demonstrated OOK line rates. Energy-efficient and fast modulators are found in the lower right corner of the diagram. With our **SOH MZM**, we generate line rates of up to 100 Gbit/s at drive voltages as low as 1.4 V_{pp} (★), corresponding to a switching energy of 98 fJ/bit. This marks the highest OOK line rate reported so far at the lowest drive voltage. A previous OOK data transmission [20] experiment at 40 Gbit/s with a drive voltage of only 950 mV is also given as a reference.

The fastest and most energy-efficient competing devices are based on InP, where both electro-absorption modulators (EAM) and MZM have been shown. **InP EAM** enable line rates up to 80 Gbit/s at peak-to-peak drive voltages of 3 V_{pp} [108]. With an **InP MZM**, 60 Gbit/s OOK signaling was demonstrated at a comparatively low drive voltage of 1.5 V [109]. More recently 100 Gbit/s at drive voltages of 2.3 V_{pp} were shown [110]. This is clearly higher than the 1.4 V_{pp} demonstrated for our devices, and the associated switching energy of more than 250 fJ/bit is more than twice the value reported for our device. Further, 100 Gbit/s OOK was demonstrated [111] with an InP MZM at a peak-to-peak drive voltage of 1.5 V_{pp}. This performance is on par with the results obtained by our experiment, however the device features a π-voltage of 2 V which is twice as high as the 0.9 V obtained for the SOH

MZM. This allowed to drive the modulator in the linear regime for multi-level PAM signaling but sacrificed modulation depth. Moreover, SOH devices exploit the intrinsic scalability advantages associated with large-scale silicon photonic fabrication based on highly developed CMOS processes, potentially in combination with electrical circuits [62], [142]. Regarding the silicon photonic platform, silicon-germanium (**SiGe**) **EAM** were realized and demonstrated at line rates of up to 50 Gbit/s, using drive voltages of at least 2 V_{pp} [112], [113]. More recently, a 100 Gbit/s OOK signal was generated using a SiGe EAM [114] at a drive voltage of 2 V_{pp}. With all-silicon MZM (**Si MZM**), line rates up to 70 Gbit/s [115], 80 Gbit/s [116], and 90 Gbit/s [117] were demonstrated, but at the expense of large drive voltages of 5 V_{pp} and beyond. Another interesting approach published recently relies on a silicon-based plasmonic-organic hybrid (**POH**) **MZM**, for which a line rate of 72 Gbit/s and 100 Gbit/s have been published [102]. It shall be noted that the 72 Gbit/s signal was detected by a coherent receiver and hence signal quality aspects cannot be directly compared to the other direct detection experiments. In general, the performance of these is fundamentally limited by an intrinsic trade-off between energy consumption and insertion loss [17], leading to relatively large drive voltages of 6 V_{pp} [97] and 4 V_{pp} [102], respectively. In addition, fabrication of POH devices has so far relied on electron-beam lithography, which is not readily amenable to large-scale low-cost production. Figure 3.5 (b) takes the same MZM references but plots the π-voltages instead of the applied drive voltage. The plot nicely exhibits the extraordinary low π-voltages of less than 1 V as the unique feature of SOH modulators. For the **InP MZM**, the π-voltages amount to 1.7 V and 2 V for the devices demonstrating 60 Gbit/s and 100 Gbit/s, respectively. The π -voltage of the **POH MZM** is not specified in the paper, but estimated from the device length and from the $U_{\pi}L$ product that is referenced for POH modulators using the same EO material. Note that the EAM listed in Figure 3.5 (a) are not included in the comparison in (b), simply because the π-voltage is not defined for an EAM. Moreover, Figure 3.5 (b) does not contain all MZM from Figure 3.5 (a), since some of the specified or estimated π-voltages would exceed the scale.

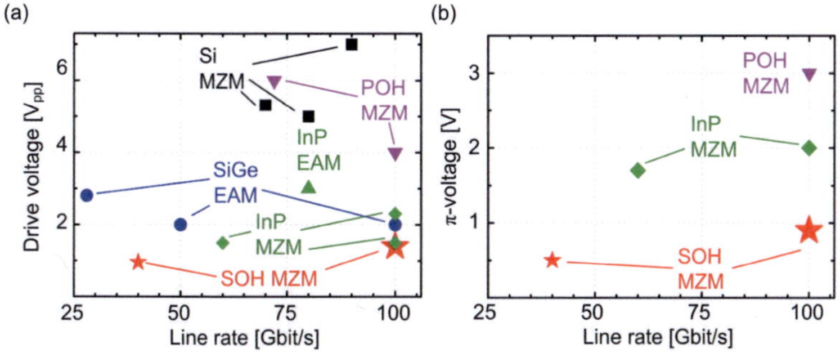

Figure 3.5: Comparison of OOK modulators fabricated on various material platforms. (a) Comparison of drive voltages and OOK line rates for different modulators types. Silicon germanium (**SiGe**) electro-absorption modulators (**EAM**), 28 Gbit/s @ 2.8 V_{pp}, 50 Gbit/s @ 2 V_{pp} [112], [113]. Indium phosphide (**InP**) **EAM**, 80 Gbit/s @ 3 V_{pp} [108]. **InP MZM**, 60 Gbit/s @ 1.5 V_{pp}, 100 Gbit/s @ 2.3 V_{pp} [109], [110] and @ 1.5 V_{pp} [111]. All-silicon Mach-Zehnder modulators (**Si MZM**) 70 Gbit/s @ 5.3 V_{pp}, 80 Gbit/s @ 5 V_{pp} [115], [116]. Plasmonic-organic hybrid (**POH**) **MZM**, 72 Gbit/s @ 6 V_{pp} [97] and 100 Gbit/s @ 4 V_{pp} [102]. Present work: Silicon-organic hybrid (**SOH**) **MZM**, 100 Gbit/s @ 1.4 V_{pp}, corresponding to a switching energy of 98 fJ/bit. This corresponds to the by far lowest value demonstrated to date. Another experiment with SOH modulators [20] was performed with drive voltages of 950 mV at a line rate of 40 Gbit/s. (b) Comparison of modulators in terms of OOK line rate and π-voltage. The **SOH MZM** for 40 Gbit/s and 100 Gbit/s exhibit π -voltages of 0.5 V and 0.9 V, respectively. These values are clearly below those achieved by other material platforms: **InP MZM** with π -voltages of 1.7 V and 2 V were operated at 60 Gbit/s and 100 Gbit/s, respectively. For the **POH MZM**, we estimate a U_π of 3 V. In the plot, EAM are not depicted since the π-voltage is not a relevant figure of merit for these devices. For the Si MZM, the π -voltage is either not given or π -voltage of more than 20 V are estimated [115], which are outside the plotted parameter range. Similarly, the π-voltage of approximately 10 V in the 72 Gbit/s POH experiment cannot be depicted.

As an example, the 70 Gbit/s **Si MZM** [115] and the 72 Gbit/s **POH MZM** [97] feature π-voltages of 20 V and around 10 V, respectively, which would be outside the depicted range. For VCSEL-based links, the highest OOK data rate demonstrated so far amounts to 70 Gbit/s [105], [143], but transmission distances are usually less than 100 m.

SOH devices can hence compare very well to competing device concepts, both in terms of performance and in terms of device scalability. The record-low drive voltage of 1.4 V_{pp} can be further reduced by using optimized poling techniques [144] or advanced organic materials, for which EO coefficients in excess of 350 pm/V have been demonstrated [72]. This might pave the path towards SOH MZM with sub-500 mV drive voltages that can be directly operated [76] by energy-efficient CMOS circuits without additional SiGe BiCMOS amplifier stages. Note that serializer/deserializer (SerDes) chips generating 100 Gbit/s NRZ drive signals have recently been realized on the BiCMOS platform [141]. SOH modulators are perfectly suited to complement these circuits on the optical side and to enable short-reach high-speed transceivers with unprecedented energy efficiency that will be key for future Ethernet interfaces at data rates of 400 Gbit/s, 800 Gbit/s, or 1.6 Tbit/s. In the future, such devices might be co-integrated with light sources in compact chip-scale assemblies. This can be accomplished by hybrid integration approaches that rely, e.g., on flip-chip integration of direct-bandgap III-V dies on processed silicon photonic waveguides [145], on mounting of readily processed III-V lasers onto silicon photonic dies [146], or on photonic multi-chip integration concepts that exploit the concept of photonic wire bonding [128], [130], [131].

Another important aspect of the SOH concept is the stability of the organic EO materials. Recent advances in material synthesis and molecular design have led to efficient and thermally stable organic EO materials. Meanwhile, tests have shown that organic EO materials [80] such as SEO100 or side-chain polymers [84] maintain more than 90% of their EO activity when stored at an elevated temperature of 85°C for 500 hours. We have previously shown that SOH modulators with SEO100 are suited for operation at elevated temperatures of 80°C under ambient atmospheric conditions [22]. More recently, polymer modulators withstanding temperatures of 105°C for 2000 hours have been demonstrated [147]. We expect that even higher operating temperatures can be achieved in the future, e.g., by cross-linking techniques [83], or by using materials with intrinsically higher glass transition temperatures such as side-chain EO polymers [84]. In addition to temperature, photo-oxidation might play an important role in the degradation of EO organic materials – this aspect is subject to ongoing research, and we expect that the associated life-time limitations can also be overcome by cross-linking, which reduces oxygen diffusion into the material [87], [88], or by encapsulation of the devices.

3.4 Conclusion

Using a silicon-organic hybrid (SOH) MZM we demonstrate for the first time the generation and transmission of a 100 Gbit/s OOK signal at record-low drive voltages of only 1.4 V_{pp} and energy consumptions of only 98 fJ/bit. We confirm our experimental results by simulations using the measured frequency characteristic of our SOH modulator and its associated drive circuitry. The moderate EOE modulator bandwidth of 25 GHz leads to an eye opening, which is much improved by the nonlinear modulator transfer characteristic. In the experiments, we achieve BER below the threshold for hard-decision forward error correction. Post-equalization helps in improving the eye opening so that low-power and low-latency FEC codes can be employed. The efficiency of our modulator and the possibility to exploit large-scale silicon photonic integration allows the realization of compact and technically simple high-speed transceivers that meet the stringent cost targets of medium-reach interconnects.

[end of paper [J19]]

4 High-speed data transmission using SOH IQ modulators

This chapter reports on the use of SOH IQ modulators for 100 GBd operation. The 16QAM modulation format leads to a line rate of 400 Gbit/s. It is taken from paper [J20] that is published *Optics Express*. In order to fit the structure and layout of this document, it was adapted accordingly.

[start of paper [J20]]

Coherent modulation up to 100 GBd 16QAM using silicon-organic hybrid (SOH) devices

Optics Express, Vol. 26, Issue 1, pp. 220-230 (2018)

DOI: 10.1364/OE.26.000220

S. Wolf,[1] H. Zwickel,[1] C. Kieninger,[1,2] M. Lauermann,[1,3] W. Hartmann,[1,2,4] Y. Kutuvantavida,[1,2] W. Freude,[1] S. Randel,[1] and C. Koos,[1,2]

[1] Karlsruhe Institute of Technology, Institute of Photonics and Quantum Electronics, 76131 Karlsruhe, Germany
[2] Karlsruhe Institute of Technology, Institute of Microstructure Technology 76344 Eggenstein-Leopoldshafen, Germany
[3] Now with: Infinera Corporation, Sunnyvale, CA, United States
[4] Now with: Physikalisches Institut, University of Muenster, 48149 Muenster, Germany

We demonstrate the generation of higher-order modulation formats using silicon-based in-phase/quadrature (IQ) modulators at symbol rates of up to 100 GBd. Our devices exploit the advantages of silicon-organic hybrid (SOH) integration, which combines silicon-on-insulator waveguides with highly efficient organic electro-optic (EO) cladding materials to enable small drive voltages and sub-millimeter device lengths. In our experiments, we use an SOH

IQ modulator with a π-voltage of 1.6 V to generate 100 GBd 16QAM signals. This is the first time that the 100 GBd mark is reached with an IQ modulator realized on a semiconductor substrate, leading to a single-polarization line rate of 400 Gbit/s. The peak-to-peak drive voltages amount to 1.5 V_{pp}, corresponding to an electrical energy dissipation in the modulator of only 25 fJ/bit.

4.1 Introduction

Advanced modulation formats such as quadrature phase-shift keying (QPSK) and 16-state quadrature amplitude modulation (16QAM) are widely deployed in metro and long-haul networks. This has significantly contributed to increase the data rates that can be transmitted over a single wavelength-division multiplexing (WDM) channel. However, it is additionally necessary to increase the underlying symbol rates, while the underlying devices must be amenable to compact and cost-efficient integration to maintain economical and technical scalability of high-performance WDM systems to large channel counts [148].

In laboratory experiments, net data rates up to 1 Tbit/s on a single optical carrier were demonstrated using QAM modulation at symbol rates up to 100 GBd [45], [103]. However, these experiments relied on conventional lithium-niobate modulators, which lack the potential for dense photonic integration and which were combined with an optical equalizer to compensate the low-pass characteristics of the device. As a promising alternative, modulators based on semiconductors like indium-phosphide (InP) or silicon (Si) have been explored, offering comparable bandwidths, but much smaller footprint. Using high-bandwidth InP modulators, the generation of single-carrier signals at line rates (net data rates) of up to 352 Gbit/s (286 Gbit/s) has been demonstrated [149], [150]. However, while InP offers the potential to monolithically integrate modulators and lasers, it requires expensive processing and is not amenable to co-integration of electronic circuits.

In contrast to that, silicon photonics [14], [151]–[154] leverages advanced large-scale cost-efficient complementary metal-oxide-semiconductor (CMOS) fabrication processes [11], [153], which are designed for high integration density and high yield, and which offer a path towards co-integration of photonic circuits with electronics [62]. However, up to now, conventional all-silicon photonic (SiP) IQ modulators still lag behind their InP counterparts, both in terms of transmission speed and efficiency. The highest reported line rate for SiP modulators amounts to 227 Gbit/s per polarization, exploiting electrical orthogonal frequency-division multiplexing (eOFDM) [155]. In this experiment, a forward-error correction (FEC) algorithm with a code rate of 0.68 (47 % overhead) was used to achieve error-free transmission after 480 km, thus leading to a net data rate of 154.2 Gbit/s. In another experiment, a net data rate of 150 Gbit/s was generated using a 30 GBd polarization-division multiplexing (PDM) 64QAM signal [14]. Hence, up to now, using SiP IQ modulators, the 400 Gbit/s mark, corresponding to a single-polarization net data rate of 200 Gbit/s, has not yet been reached. Moreover, all these devices rely on carrier depletion in reversely biased p-n junctions [13], which leads to a comparatively low modulation efficiency. In general, the modulation efficiency of electro-optic (EO) modulators is quantified by the π-voltage-length product $U_\pi L$, which typically amounts to a relatively large value of the order of 10 Vmm for depletion-type SiP MZM [67], [156], [157]. As a consequence, broadband drive amplifiers providing voltage swings between 2.5 V_{pp} and 5 V_{pp} are required to operate these devices [15], [158].

Silicon-organic hybrid (SOH) modulators [17], [159] overcome the limited efficiency of all-silicon devices by combining conventional silicon-on-insulator (SOI) waveguides with highly-efficient organic EO cladding materials [73], [124]. Typically, organic EO materials feature negligible absorption losses for wavelengths above 1200 nm [77], [79], [86] while maintaining high electro-optic activity [86], [160]. SOH devices can hence be used over a broad range of infrared wavelengths, including common telecommunication bands around 1310 nm and 1550 nm. The underlying SiP waveguide structures can be fabricated using optical lithography, and the organic materials are deposited in a back-end-of-line process. With voltage-length products down to 0.5 Vmm [20], [71], the efficiency of SOH Mach-Zehnder modulators (MZM) surpasses the efficiency of all-silicon devices by more than one order of magnitude. Using SOH MZM, on-off-keying (OOK) signals have been generated with drive voltages of only 80 mV_{pp} [20], and the devices are also well suited for higher-order modulation formats [76], [118], [119]. The generation of 112 Gbit/s

16QAM signals has been demonstrated at a voltage swing of only 600 mV_{pp} without any drive amplifiers [119]. SOH IQ modulators can also be operated directly from the binary outputs of a field-programmable gate array (FPGA), where line rates of up to 52 Gbit/s were demonstrated [76]. Besides low drive voltages, SOH EO modulators allow high-speed signaling for both coherent and direct-detection transmission [118], [121], [161].

In this paper, we demonstrate that the SOH concept allows to further push the limits of silicon-based IQ modulators by enabling 16QAM signaling at symbol rates of up to 100 GBd. In a first set of experiments, we use CMOS digital-to-analog converters (DAC) to drive an SOH IQ at 16QAM symbol rates of 63 GBd, thus leading to a single-polarization line rate of 252 Gbit/s [162]. The bit error ratio (BER) amounts to 4.1×10^{-3} such that forward error correction (FEC) schemes with a 7 % overhead can be applied, thus leading to an error-free reception with a net data rate of 232 Gbit/s. In a second set of experiments, we use BiCMOS DAC and employ a bandwidth-increasing gate field [19], [92], [94] to demonstrate 16QAM signaling at line rates of up to 100 GBd, corresponding to single-polarization line rates of up to 400 Gbit/s. For the 100 GBd 16QAM demonstration, we obtain BER values that are below the limits of state-of-the-art third-generation FEC algorithms. Considering a code rate of 0.83 (20 % overhead), the line rate of 400 Gbit/s corresponds to a net data rate of more than 330 Gbit/s [123]. This represents the highest line rate and the highest net data rate so far demonstrated for an integrated EO modulator fabricated on a semiconductor substrate. Exploiting the second polarization would enable single-carrier line rates of up to 800 Gbit/s. The 63 GBd and 100 GBd signals were generated with small peak-to-peak drive voltages of 1.0 V_{pp} and 1.5 V_{pp}, respectively – the lowest value so far demonstrated for semiconductor-based IQ modulators at these speeds. The electrical energy consumption of the modulator amounts to only 25 fJ/bit for the 400 Gbit/s signal – more than an order of magnitude below that of conventional all-silicon modulators at considerably lower speeds [15], [158]. These proof-of-principle experiments underline the outstanding properties of SOH modulators and show a path towards coherent silicon-based transmitters operating at line rates of 400 Gbit/s and above without the need for power-hungry drive amplifiers.

4.2 Silicon-Organic Hybrid (SOH) IQ Modulator

Silicon-organic hybrid (SOH) modulators rely on the interaction of the optical wave with an EO organic cladding material in a silicon-on-insulator (SOI) slot waveguide [17]. The concept of an SOH IQ modulator is depicted in Figure 4.1 (a). The IQ modulator consists of two nested Mach-Zehnder (MZ) interferometers formed by SOI waveguides and multi-mode interference (MMI) couplers, which split or combine the fields at their inputs. A schematic of the cross-section A - A' of an SOH MZM is depicted in Figure 4.1 (b). The phase shifter sections are realized as SOI slot waveguides which are formed by two 240-nm wide silicon (Si) rails that are separated by the 120-nm wide slots.

The slots are filled with an organic electro-optic cladding material. For the fundamental quasi-TE mode (dominant transverse electric field component parallel to the substrate), the high refractive-index contrast between waveguide and EO cladding leads to a strong confinement inside the slot [70], as visualized in Figure 4.1 (c). Utilizing the high linear electro-optic coefficient of the organic cladding material, the refractive index is modulated by an electric field between the two silicon rails. The modulating electric field is generated by applying a drive voltage, $U_{\mathrm{m}}(t)$, to coplanar ground-signal-ground (GSG) travelling-wave electrodes, which are connected to the silicon rails via 70-nm high slightly *n*-doped conductive slabs, see Figure 4.1 (b). This design ensures that the modulating electric field $E_{x,RF}(t)$ is also confined in the narrow slot, Figure 4.1 (d), thus leading to a strong overlap of the optical and the modulating electric fields and resulting in a high modulation efficiency with $U_{\pi}L$ products down to 0.5 Vmm [20]. In contrast to previous demonstrations of SOH devices [22], [118], which still relied on electron beam lithography for structuring the slot waveguides, the devices in this work are fabricated using the 248 nm deep-UV (DUV) process at A*Star IME [163], which allows for highly reproducible large-scale processing at high yield and low cost and for co-integration of SOH devices with the full portfolio of standard silicon-photonic components such as Germanium photodiodes and thermal phase shifters. Fabrication at IME required some adaptations of the modulator and the transmission line design to comply with the design rules.

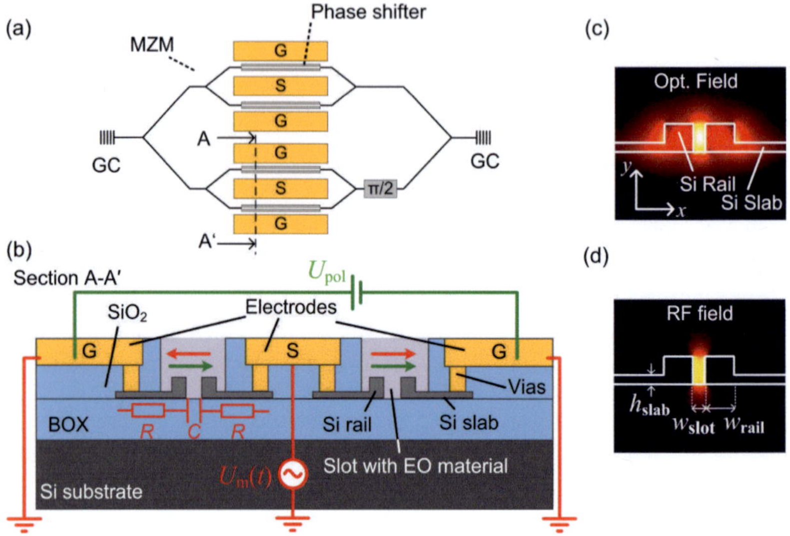

Figure 4.1: Concept of the silicon-organic hybrid (SOH) modulator. (a) Schematic of an SOH in-phase/quadrature (IQ) modulator consisting of two nested Mach-Zehnder modulators (MZM). The thin black lines represent standard silicon strip waveguides. The phase modulator sections are based on slot waveguides (rail width $w_{rail} = 240$ nm, slot width $w_{slot} = 120$ nm) and represented by slotted rectangles in light grey. Coplanar ground-signal-ground (GSG) radio-frequency transmission line carry the modulation signals. (b) Schematic of a SOH MZM cross section along the line AA' illustrated in (a). The transmission line is electrically connected to the slot by aluminum (Al) vias and thin n-doped silicon slabs (thickness $h_{slab} = 70$ nm, width 8 µm). The chip is overclad with SiO_2 which is locally removed in the slot areas. The slots are covered with an organic electro-optic (EO) material, which is deposited on the chip such that it homogeneously fills the slots. The chromophores are aligned (green arrows) at an elevated temperature using a poling voltage U_{pol} applied to the floating ground (G) electrodes. After cooling to ambient temperature, the orientation of the chromophores is frozen, and the poling voltage can be removed. For operation of the device, the modulating signal is applied to the GSG line. This RF field (red arrows) is oriented in opposite direction to the chromophore alignment in one arm, and in the same direction in the other arm of the MZM so that the two phase modulators of the MZM are operated in push-pull mode. (c) Fundamental quasi-TE mode of the slot waveguide. The color-coded graph shows the magnitude of the dominant optical electric field (dominant x-component), which is strongly confined to the slot. (d) Magnitude of the electrical modulation field (dominant x-component) which is also strongly confined to the slot. The good overlap of optical and modulating fields results in efficient EO modulation.

Note that Figure 4.1 (b) gives only a simplified schematic of the structure – in the real device, the silicon slot waveguides, the actually buried metal transmission line, and the contact pads at the chip surface are connected by aluminum (Al) vias. In Figure 4.1 (b), only the transmission line metal layer is depicted. For the deposition of the EO material, the slots are accessed by locally opening the oxide cladding in the respective regions. In contrast to earlier demonstrations of SOH devices, we have developed dedicated dispensing processes that allow for locally depositing the EO material instead of applying it to the entire chip by spin coating. In order to achieve EO activity in the cladding, the material is poled in a one-time procedure after deposition. To this end, the device is first heated close to its glass-transition temperature. A poling voltage U_{pol} is then applied across the floating ground electrodes, see Figure 4.1 (b), such that half of the voltage drops across each of the slots. This leads to an alignment of the dipolar chromophores in the slot in the direction of the poling field, indicated by the green arrows in Figure 4.1 (b). Finally, the device is cooled down to ambient temperatures while holding the poling voltage. The chromophore alignment is thus "frozen" and will remain, even after removing the poling voltage. Applying the modulating voltage $U_{\mathrm{m}}(t)$ to the GSG electrodes then leads to electric fields, indicated by red arrows, Figure 4.1 (b). These fields are oriented in the same (opposite) direction as the chromophores in the right (left) slot. Consequently, the phase shifts in both arms have equal magnitude but opposite sign, and the SOH MZM operates in push-pull mode when applying a single-ended drive signal, allowing nearly chirp-free modulation [161].

The modulation bandwidth of SOH modulators is limited by an *RC* characteristic resulting from the resistivity of the silicon slabs (represented by a lumped resistor *R*) and the capacitance of the slot (represented by a lumped capacitance *C*). Previous high-speed experiments with SOH modulators [22], [118], [121] relied on the use of a gate voltage applied between the bulk silicon and the device layer. The gate voltage leads to a highly conductive electron accumulation below the Si slab surface near the SiO_2-layer, which decreases the slab resistivity and consequently increases the modulation bandwidth [19], [92]. In future designs, an improved lateral doping profile of the 8-µm wide silicon slabs could eliminate the need for a gate voltage without compromising the optical attenuation.

In the following sections, we describe two sets of high-speed modulation experiments, obtained with and without a gate voltage. In the first set of experiments, we show single-polarization 16QAM modulation and detection at symbol rates up to 63 GBd, resulting in line rates (net data rates) of up to 252 Gbit/s (235 Gbit/s) at an energy consumption of down to 22 fJ/bit. These experiments were performed without a gate field. In a second set of experiments, we use a gate voltage to push the modulator limits, breaking the 100 GBd mark. This leads to single-polarization 16QAM optical signals with line rates (net data rates) up to 400 Gbit/s (333 Gbit/s), generated at a slightly higher energy consumption of 25 fJ/bit.

4.3 Setup for optical data generation

The basic experimental setup is depicted in Figure 4.2. The electrical drive signals are derived from an arbitrary-waveform generator (AWG) and rely on random bit sequences. Radio-frequency (RF) amplifiers are used to increase drive voltage levels. The GSG transmission lines of the SOH IQ modulator are fed using a microwave probe and terminated by external 50 Ω resistors that are connected via a second probe. The optical carrier is provided by an external cavity laser (ECL) and coupled to and off the chip via grating couplers (GC). An erbium-doped fiber amplifier (EDFA) after the modulator compensates the insertion loss and acts as a receiver pre-amplifier, and an optical bandpass filter is used to remove out-of-band amplified spontaneous emission (ASE) noise. The signal is received back-to-back by an optical modulation analyzer (OMA) and evaluated off-line. In the experiments, the length of the signal recording is limited by the available hardware. The bit error ratio (BER) can be measured directly in case enough errors are found within the signal recordings. Otherwise, we use the error vector magnitude (EVM) [51] as quality metric. The EVM describes the deviation of a received constellation point from its ideal position in the complex plane and can be used to estimate the BER under the assumption that the signal is distorted by additive white Gaussian noise only [36], [51], [52].

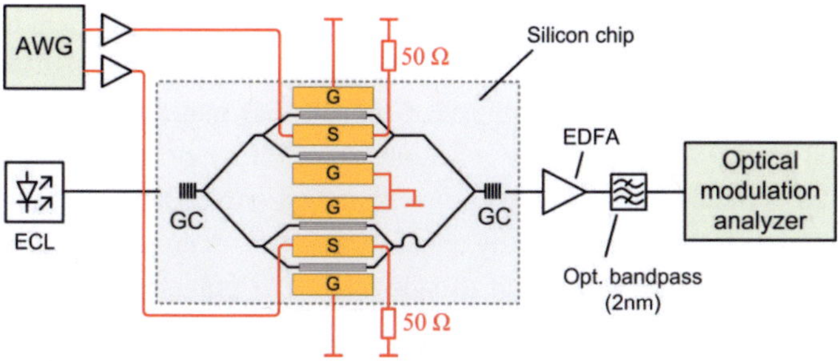

Figure 4.2: Experimental setup. An arbitrary-waveform generator (AWG) is used for driving the modulator via radio-frequency (RF) amplifiers. The electrical drive signals are fed to the ground-signal-ground (GSG) transmission lines of the SOH IQ modulator. The transmission lines are terminated by external 50 Ω resistors. The optical carrier is provided by an external cavity laser (ECL). Grating couplers (GC) are used to couple light to and off the chip. The insertion loss of the modulator is compensated by an erbium-doped fiber amplifier (EDFA), and an optical bandbass filter is used to suppress out-of-band amplified spontaneous emission (ASE) noise. The signal is received, recorded and evaluated using an optical modulation analyzer (OMA). We perform two sets of experiments, which we refer to as Experiment 1 and Experiment 2 in the following. Both experiments use the same setup scheme, but differ in the specifications of the actual equipment, in particular in the bandwidth of the AWG and the OMA, see Appendix F for details.

Using the scheme depicted in Figure 4.2, we performed two sets of experiments, which are referred to as Experiment 1 and Experiment 2 in the following. Both experiments use the same setup scheme, but differ in the specifications of the actual equipment, in particular in the analog bandwidth of the electronics at the transmitter and receiver, see Appendix F for a detailed list and the technical specifications of the various equipment items. The two investigated SOH modulators have virtually the same specifications: In the first (second) experiment, the phase shifter length of the IQ modulator amounts to 0.5 mm (0.6 mm), and $U_\pi L$-products of 1.1 Vmm (1.0 Vmm) were achieved by using SEO100 (SEO250) as EO material [78]–[80], [82]. Note that these $U_\pi L$-products are slightly larger than the ones reported in previous work [20], [71] because the EO materials were chosen for good temperature stability rather than for highest EO activity. For SEO100, EO coefficients of $r_{33} = 166$ pm/V have been achieved in bulk material, and the glass transition temperature amounts to 140 °C [79]. Using this material, we have previously demonstrated device operation at temperatures of 80 °C [22]. We expect that the operating temperatures and long-term stability of organic EO materials can be further increased in the future, e.g., by reducing the mobility of the chromophores, which helps maintaining the acentric alignment established during poling. This can, e.g., be accomplished by using materials with an intrinsically higher glass transition temperature such as side-chain EO polymers [84], or by exploiting cross-linkable materials [83], in which additional chemical bonds are established after poling. In the first experiment, the insertion loss of the SOH IQ modulator chip amounts to approximately 17.5 dB, dominated by the fiber-chip coupling losses of approximately 9 dB (4.5 dB per GC-interface). The device used in the second experiment had a higher insertion loss due to a specific chip design which contained additional power splitters in the feeding waveguide sections. We expect that the total fiber-to-fiber insertion losses (on-chip losses) of the IQ modulator can be reduced to significantly less than 10 dB (5 dB) by improved fabrication processes that reduce sidewall roughness, by asymmetric slot waveguide geometries [164], and by optimized doping profiles for the phase shifter sections [118].

4.4 Signal generation up to 400 Gbit/s

In Experiment 1, we use an AWG featuring an analog bandwidth of 32 GHz and a sampling rate of 92 GSa/s (Keysight M8196A) along with an OMA featuring an analog bandwidth of 63 GHz and a sampling rate of 160 GSa/s (Keysight DSOZ634A). We first generate single-polarization QPSK signals with data rates up to 154 Gbit/s, corresponding to symbol rates of up to 77 GBd – the corresponding constellation diagrams are shown in Figure 4.3 (a). For these signals, a direct measurement of the BER was not possible since no errors were measured in our recordings of 6.5×10^6 bit. Instead, we used the measured EVM values to estimate the associated BER_e values, assuming that the signal is distorted by additive white Gaussian noise only [36], [51]. The measured EVM values of 11.3 % and 13.6 % for symbol rates of 46 GBd and 56 GBd correspond to estimated BER_e values well below 10^{-15} and 10^{-12}, respectively.

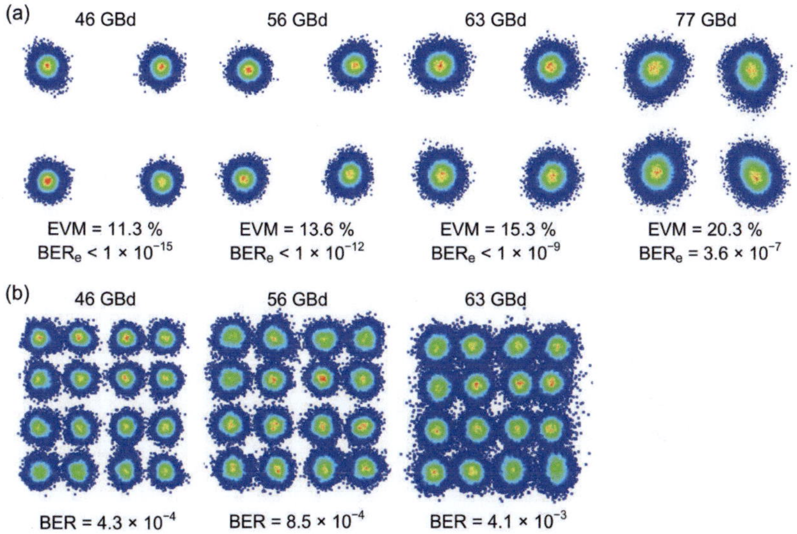

Figure 4.3: Results of Experiment 1. Constellation diagrams for single-polarization (a) QPSK and (b) 16QAM signals with symbol rates of up to 63 GBd for 16QAM signaling and up to 77 GBd for QPSK. In the QPSK experiment, we measure the error vector magnitude (EVM) and estimate a corresponding bit error ratio (BER_{est}). With estimated BER values below 10^{-9}, the QPSK signal can be considered error free for symbol rates up to 63 GBd. For the 77 GBd QPSK signal with a line rate of 154 Gbit/s on a single polarization, we measure an EVM of 20.3 % corresponding to an estimated BER of 3.6×10^{-7} – still well below the threshold of hard-decision FEC with 7 % overhead. For the 16QAM signals, the BER is directly measured from the signal recordings. The BER values range from 4.3×10^{-4} for the 46 GBd signal to 4.1×10^{-3} for the 63 GBd signal representing a line rate of 252 Gbit/s. All 16QAM BER values are hence well within the limits of hard-decision FEC with 7 % overhead.

The 63 GBd signal can still be considered error-free as we estimate a $BER_e < 10^{-9}$ from the EVM of 15.3%. For symbol rates of 73 GBd and 77 GBd, the measured EVM of 19.7% and 20.3% correspond to BER_e values of 2×10^{-7} and 3.6×10^{-7}, respectively. A direct measurement of the BER was not possible for the QPSK signals as no errors were measured in our recording of 6.5×10^{6} bits. Still, these values are clearly below the 4.5×10^{-3} limit for hard-decision forward error correction (FEC) codings [133] with 7 % overhead. The received signals were evaluated using Keysight's vector signal analysis (VSA) software [165] using a series of digital processing stages such as low-pass filtering, polarization demultiplexing, chromatic dispersion compensation, frequency offset estimation, carrier phase estimation and adaptive equalization.

The (single-polarization) 16QAM constellation diagrams are depicted in Figure 4.3 (b) – here we measure BER values directly. The lowest BER value is 8×10^{-6} for a 23 GBd 16QAM signal corresponding to a line rate of 92 Gbit/s (not depicted). For the 46 GBd and 56 GBd (184 Gbit/s and 224 Gbit/s) signals, the measured BER is 4.3×10^{-4} and 8.5×10^{-4}, respectively. For the highest symbol rate of 63 GBd, corresponding to a line rate of 252 Gbit/s for a 16QAM modulation in a single polarization, the BER is 4.1×10^{-3}. All of the measured BER values are within the 4.5×10^{-3} limit for hard-decision FEC coding with 7 % overhead. Considering the FEC overhead, the single-polarization line rates of 154 Gbit/s for the 77 GBd QPSK signal and of 252 Gbit/s for the 63 GBd 16QAM signal correspond to net data rates of 143 Gbit/s and 234 Gbit/s, respectively.

In Experiment 2, we expand the analog bandwidths of both the AWG and the OMA for increasing the symbol rate of the 16QAM signaling. As an AWG, we use a Micram DAC4 featuring an analog bandwidth of more than 40 GHz and a sampling rate of up to 100 GSa/s [166]. The OMA comprises a coherent receiver (Tektronix OM4245) featuring an analog bandwidth of 45 GHz and 70 GHz real-time oscilloscopes with sampling rates of 200 GSa/s (Tektronix DPO77002SX). We further apply a gate field [19], [92], [94] of 0.1 V/nm to increase the modulator bandwidth. The larger transmitter bandwidth allows for symbol rates up to 100 GBd at the cost of a higher noise level.

This requires advanced FEC schemes with BER thresholds of 2×10^{-2} or even higher, which may use soft-decision or staircase architectures [167]–[169]. We further use dedicated digital signal processing (DSP) at the receiver, including digital timing recovery along with a 33-tap fractionally spaced (two-fold oversampled) adaptive feed-forward equalizer that is adapted by a least-mean-square stochastic gradient algorithm [170]. To this end, we use training symbols to initialize the coefficients, before we switch to a decision-directed operation. We compute the BER from a sample of more than 500 000 bit.

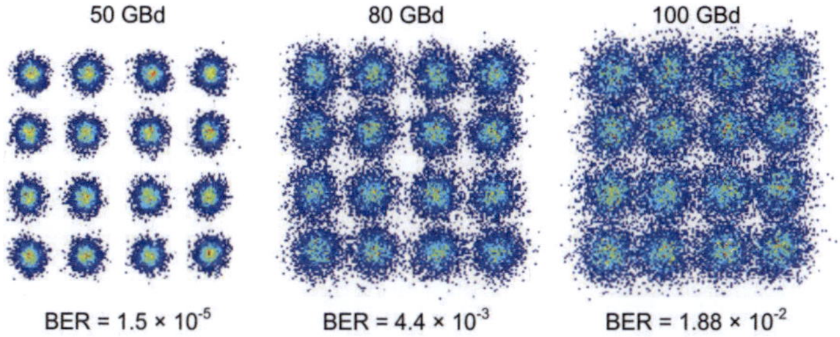

Figure 4.4: Results of Experiment 2: 16QAM constellation diagrams for symbol rates of 50 GBd (left), 80 GBd (center) and 100 GBd (right). The BER up to 80 GBd (320 Gbit/s) remains below the hard-decision FEC limit with 7 % overhead, while the 100 GBd (400 Gbit/s) signal is below the soft-decision FEC threshold with 20 % overhead. The single-polarization net data rate is 333 Gbit/s.

The results for the single-polarization 16QAM generation are depicted in Figure 4.4. At 50 GBd (200 Gbit/s) and 80 GBd (320 Gbit/s), we measure a BER of 1.3×10^{-5} and 3.8×10^{-3}, respectively. The measurement shows an improvement with respect to the previous results, enabled by the bandwidth improvements in the setup and by the optimized DSP. Both BER values are within the threshold of a hard-decision FEC with 7 % overhead, leading to net data rates of 186 Gbit/s and 299 Gbit/s. At 100 GBd, corresponding to a line rate of 400 Gbit/s on a single polarization, the measured BER of 1.7×10^{-2} is within the limits of today's soft-decision FEC codes. The optical signal-to-noise power ratio (OSNR) measured for a spectral bandwidth of 0.1 nm amounts to 27.5 dB, which corresponds to an implementation penalty of Δ OSNR = 8.5 dB for the single-polarization 16QAM signal at 100 GBd with respect to the theoretical value [48]. We attribute this implementation penalty to bandwidth limitations of the RF components and of the modulator and to the fact that we did not pre-equalize the drive signals. Considering a FEC overhead of 20%, this results in a net data rate of more than 330 Gbit/s. Note that, in contrast to Experiment 1, the 100 GBd signals are generated without oversampling such that pulse shaping techniques cannot be used. For a fair comparison of the results, the 50 GBd signal (2-fold oversampling) and the 80 GBd signal (no oversampling) are also generated without pulse shaping. To the best of our knowledge, our experiments represent the first demonstration of 100 GBd signaling using an IQ modulator realized on a semiconductor substrate.

4.5 Energy considerations

Power dissipation is an important aspect in today's optical communication networks. Especially the electronic drivers for EO modulators contribute significantly to the total power consumption of a transceiver unit. The specifications for the drivers are dictated by the choice of the EO modulator and the required drive voltages. As a figure of merit which is related to the transceiver's energy consumption, we use the power dissipation in the modulator.

For an estimate of the IQ modulator power dissipation, we assume that the transmission line impedances of the two MZM are perfectly matched to the terminating resistances $R_t = 50\ \Omega$ and to the $50\ \Omega$ internal resistances of the driver. We further assume that all symbols of the 16QAM-constellation occur with equal probability. The energy per bit then depends on the line rate r and the peak-to-peak drive voltage U_m as measured from the electrical eye opening [119],

$$W_{\text{bit}}^{(16\text{QAM})} = 2\left[\frac{1}{2}\left(\frac{U_\text{m}}{2}\right)^2 + \frac{1}{2}\left(\frac{1}{3}\frac{U_\text{m}}{2}\right)^2\right]\frac{1}{R_\text{t}} \times \frac{1}{r} \tag{4.1}$$

The 100 GBd 16QAM signal with a line rate of 400 Gbit/s was generated with a peak-to-peak drive voltage of $U_m = 1.5\ V_{pp}$, the energy per bit is 25 fJ/bit. For the 252 Gbit/s experiment with a lower voltage swing of $U_m = 1\ V_{pp}$, we find a slightly reduced energy consumption of 22 fJ/bit.

4.6 Summary

We demonstrate high-speed coherent signaling using silicon-organic hybrid (SOH) IQ modulators. The SOH approach expands the capabilities of the silicon photonic integration platform by combination with highly efficient organic electro-optic (EO) materials, thereby enabling highly efficient high-speed optical phase modulation without amplitude-phase coupling. In our experiments, we use SOH modulators with π-voltage-length products of approximately $U_\pi L = 1$ Vmm to generate data streams with 16QAM modulation at line rates of up to 400 Gbit/s in a single polarization. Up to line rates of 320 Gbit/s (80 GBd), the measured BER remains within the limits of hard-decision forward-error correction with 7% overhead, resulting in a net data rate of 299 Gbit/s. At 400 Gbit/s for a 16QAM symbol rate of 100 GBd, the BER is still below the threshold for soft-decision FEC with 20% overhead, corresponding to a net data rate of 333 Gbit/s. Our experiments demonstrate the highest symbol rates and the highest data rates reported so far for IQ modulators that are realized on a semiconductor platform, thereby emphasizing the unique advantages of the SOH integration for high-speed energy-efficient coherent communications.

[end of paper[J20]]

5 DAC-less and amplifier-less signal generation using SOH modulators

In this chapter, we highlight the efficiency of SOH modulators and demonstrate that they can be operated directly from a Field-programmable gate array (FPGA) without intermediate driver amplifiers. 16QAM signals are generated without digital-to-analog converters (DAC) but simply through a passive radio frequency (RF) network. *The following chapter is taken from paper [J7]. It was modified to fit the structure and notation of this document.*

[start of paper [J7]]

DAC-less Amplifier-less Generation and Transmission of QAM Signals Using Sub-Volt Silicon-organic Hybrid Modulators

Journal of Lightwave Technology, Vol. 33, Issue 7, pp. 1425-1432 (2015)
DOI: 10.1109/JLT.2015.2394511

S. Wolf,[1] M. Lauermann,[1] P. Schindler,[1,2] G. Ronniger,[1]
K. Geistert,[1] R. Palmer,[1,3] S. Köber,[1,4] W. Bogaerts,[5,6] J. Leuthold,[1,7]
W. Freude,[1] and C. Koos[1,8]

[1] Karlsruhe Institute of Technology, Institute of Photonics and Quantum Electronics, 76133 Karlsruhe, Germany

[2] Now with: Infinera Corporation, Sunnyvale, CA, United States

[3] Now with: Coriant GmbH, 81541 Munich, Germany

[4] Now with: Chemistry Department, University of Cologne, Cologne, Germany

[5] Photonics Research Group, Department of Information Technology, Ghent University, IMEC, 9000 Gent, Belgium

[6] Now also with Luceda Photonics, 9200 Dendermonde, Belgium

[7] Now with: Swiss Federal Institute of Technology (ETH), 8092 Zurich, Switzerland

[8] Institute of Microstructure Technology, Karlsruhe Institute of Technology, 76344 Eggenstein-Leopoldshafen, Germany

We demonstrate generation and transmission of optical signals by directly interfacing highly efficient silicon- organic hybrid (SOH) modulators to binary output ports of a field-programmable gate array. Using an SOH Mach-Zehnder modulator (MZM) and an SOH IQ modulator we generate ON-OFF-keying and binary phase-shift keying signals as well as quadrature phase-shift keying and 16-state quadrature amplitude modulation (16QAM) formats. Peak-to-peak voltages amount to only 0.27 V_{pp} for driving the MZM and 0.41 V_{pp} for the IQ modulator. Neither digital-to-analog converters nor drive amplifiers are required, and the RF energy consumption in the modulator amounts to record-low 18 fJ/bit for 16QAM signaling.

5.1 Introduction

High-capacity optical interconnects are key to overcome transmission bottlenecks in information-processing systems. In this context, intimate co-integration and direct interfacing of photonic and electronic circuitry [62] is indispensable to realize low-cost transceiver modules with high throughput, low power consumption, and the capability of using wavelength division multiplexing. At the same time, optimized spectral efficiency (SE) becomes increasingly important not only for telecommunications but also for optical interconnects in data centers, where space for fiber bundles is a precious resource and where installed fibers represent the most costly part of the communication system. This calls for higher-order modulation formats [43] such as quadrature phase-shift keying (QPSK) or M-ary quadrature amplitude modulation (M-QAM). However, the associated technical complexity, cost and power consumption of today's photonic-electronic interfaces is still prohibitive for optical interconnects: The generation of higher-order modulation formats currently relies either on high-speed digital-to-analog converters (DAC) and linear electrical amplifiers or other dedicated electronic circuitry to generate the drive signals for electro-optic modulators. The energy consumption of DACs and amplifiers is considerable: Operating a 3-bit power DAC [171] at a peak-to-peak voltage of 3.2 V and a symbol rate of 42 GBd requires a total power of 2 W, which corresponds to an energy consumption of 2 W / (42 GBd × 3 bit) = 16 pJ / bit. This is already in the order of the total package power dissipation of small form-factor pluggables (SFP) or 120 Gbit/s active optical cables (AOC) that are based on directly modulated vertical-cavity surface-emitting lasers (VCSEL). Power-efficient DAC-less amplifier-less transmitters are therefore indispensable for realizing high-speed optical interconnects that exploit co-integration of photonics with digital CMOS electronics for generating advanced modulation formats. In the context of this paper, the terms "DAC-less" and "amplifier-less" describe a scheme in which binary output ports of a digital CMOS circuit are connected to an optical modulator solely via a passive electrical network that generates multi-level drive signals without the use of further active elements.

For conventional on-off-keying (OOK), amplifier-less interfaces between the outputs of field-programmable gate arrays (FPGA) and silicon-photonic micro-ring modulators were demonstrated at line rates of 2.5 Gbit/s and drive voltages of 1.2 V_{pp} [172]. The comparatively low drive voltage was enabled by the resonant behavior of the micro-ring modulator, which allows overcoming the rather low efficiency of depletion-type modulators. However, this concept cannot be easily transferred to the generation of higher-order modulation formats where the phase needs to be controlled independently of the amplitude. Moreover, the scheme requires to carefully match the ring resonance to the carrier wavelength by, e.g., heaters, which increase the power consumption considerably. Regarding advanced modulation formats, DAC-less generation of 16QAM optical signals was proposed and demonstrated by using serial or parallel configurations of multiple Mach-Zehnder modulators (MZM), each operated with binary drive signals [173]–[175]. However, this approach still requires electrical amplifiers to drive the various MZMs with peak-to-peak voltages of typically more than 2 V [175]. More recently, generation of a 16QAM optical signal from binary electrical drive signals was demonstrated using IQ modulators that consist of MZMs with segmented phase shifters [176]. However, in this experiment, an additional electrical driver circuit was needed to interface the analog drive inputs of the modulator to the digital outputs of the data source. This leads to an additional electrical power consumption of approximately 1 W, corresponding to an energy consumption of 18 pJ / bit at a data rate of 56 Gbit/s.

In this paper, we report on proof-of-concept experiments demonstrating that non-resonant and yet highly-efficient silicon-based modulators that can be operated directly with binary sub-Volt output signals of standard field-programmable gate arrays (FPGA) without the need for digital-to-analog converters or drive amplifiers [177], [178]. These devices are realized on the silicon-organic hybrid (SOH) platform [18], [71], [179], which enables generation of a wide variety of modulation formats at lowest energy consumption [20], [119], [180], [181]. We demonstrate the viability of the approach in a series of experiments: Using SOH MZM and IQ modulators, we show generation and transmission of on-off-keying (OOK) and binary phase shift keying (BPSK) signals at 10 Gbit/s as well as QPSK and 16QAM signaling at 13 GBd. For the 16 QAM experiment, we use peak-to-peak drive voltages of only 0.41 V_{pp}, leading to an ultra-low RF energy consumption of only 18 fJ/bit in the modulator – far below that of comparable all-silicon devices. We believe that DAC-less and amplifier-less interfaces between digital circuitry and highly efficient SOH modulators can enable a novel class of highly scalable optical interconnects with unprecedented energy efficiency.

The paper is organized as follows: Section 5.2 explains specifics of the SOH technology platform and describes the structure of the modulators. Experiments with an FPGA-driven Mach-Zehnder modulator using 10 Gbit/s OOK and BPSK formats are presented in Section 5.3. In Section 5.4, the DAC-less generation of QPSK and 16QAM signals is demonstrated, and data transmission over a distance of up to 100 km with a data rate of up to 52 Gbit/s is shown. Finally, we discuss the RF energy requirements of our modulators in Section 5.5.

5.2 Silicon organic hybrid (SOH) electro-optic modulators

Our silicon-organic hybrid (SOH) modulators combine silicon-on-insulator (SOI) slot waveguides with organic electro-optic (EO) claddings [94], [179]. The optical slot waveguide has two silicon "rails" [179], Figure 5.1 (a). The space between them is filled with an electro-optic organic material.

For quasi-TE polarization, the dominant field component E_x in the slot is strongly enhanced due to the discontinuity of the normal electric field component at the interface to the silicon rails. If a voltage is applied across the slot, the refractive index changes and the phase of the optical wave is modulated. Two such phase shifter sections form the arms of an SOH Mach-Zehnder interferometer as illustrated in Figure 5.1 (a).

By controlling the phase shift in both arms in a push-pull configuration, the light amplitude is modulated without introducing a residual phase shift. For modulation at microwave frequencies, copper electrodes in form of a ground-signal-ground (GSG) transmission line are added. The metal electrodes are connected to the slot by 900 nm high tungsten vias and thin n-doped silicon slabs, so that the modulating voltage drops mainly across the narrow slot of each phase shifter section. A cross-section of an SOH phase shifter together with the optical and RF field is depicted in Figure 5.1 (b) and (c), respectively.

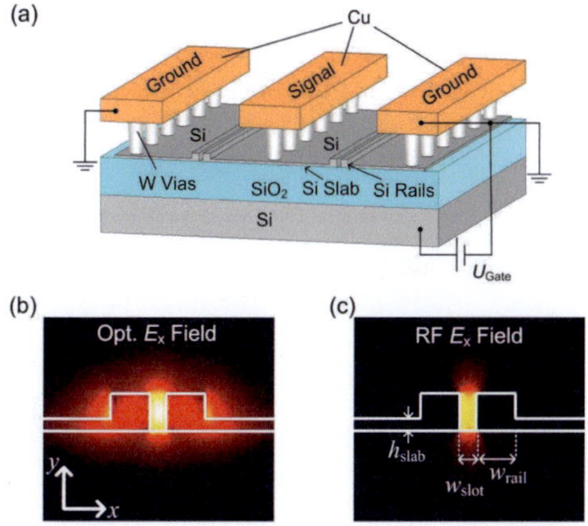

Figure 5.1: Schematic of an SOH MZM. The device consists of two phase modulators driven in push-pull operation by a single coplanar transmission line (ground-signal-ground, GSG). The GSG transmission line is made of copper (Cu) strips that are connected to the Si rails of the slot waveguide 900 nm high tungsten (W) vias and thin *n*-doped silicon slabs. The silicon slot waveguide ($w_{slot} = 120$ nm, $w_{rail} = 240$ nm, $h_{rail} = 220$ nm) is coated with the electro-optic material SEO100 (not shown). (b) E_x component of the optical field in the slot waveguide. (c) E_x component of the electrical RF drive signal. The optical mode and the modulation field are both confined to the slot, resulting in strong interaction and hence in efficient modulation.

The optical field strongly overlaps with the modulating RF field in the slot, leading to highly-efficient electro-optic interaction and hence low drive voltage requirements.

The underlying SOI structures were fabricated in IMEC's active silicon photonics platform, which includes three etch steps for the silicon waveguides as well as implants and silicidation for the electrically contacted devices. The EO material is deposited on the pre-processed device by spin-coating and fills the slot homogeneously. A separate step [71] is necessary for poling the electro-optic cladding: At an elevated temperature, a poling voltage across the (floating) ground electrodes aligns the EO chromophores in the same direction for both slots. If a modulating RF voltage is applied to the signal electrode, the associated field is oriented in opposite directions in the two arms of the MZM, leading to phase shifts of the same magnitude but opposite sign and hence to efficient push-pull operation [179]. As an EO cladding material we use SEO100 from Soluxra, LLC. The material is specified for a high electro-optic coefficient of 110 pm/V and operating temperatures of up to 85°C.

The electrical bandwidth of the modulators is limited by RC low-pass characteristics arising from the capacitance of the slot which has to be charged and discharged via the resistive silicon slabs [94]. To increase the modulation bandwidth, a static gate field of 0.1 V/nm is applied between the bulk silicon and the SOI device layer. The gate field leads to an electron accumulation layer and to a higher conductivity of the silicon slabs and hence to an increased bandwidth of the devices [19], [92], [94]. The gate voltage does not have any appreciable effect on the power consumption: The associated current flow is of the order of a few nanoamperes, corresponding to an energy consumption of a few aJ/bit, which can safely be neglected. In future device generations, the gate voltage may be reduced significantly by depositing a thin silicon layer on top of the silicon slabs [94] or even be omitted by employing an improved doping profile which increases the conductivity of the silicon slabs.

The π-voltage-length product of the modulators is below 0.8 Vmm. The π-voltages have been measured at bias voltages of 2 V or more in order to prevent screening of the applied fields by free charges in the cladding, which may occur at zero or small bias.

Due to the low mobility of the free charges, this effect is only observable for low frequencies and does not impede high-frequency operation [71].

5.3 Binary driven SOH Mach-Zehnder modulator without electrical amplification

In a first set of experiments, we use a single electrical drive signal which is fed to an MZM to generate OOK and BPSK signals. The experimental setup is depicted in Figure 5.2. Light from an external-cavity laser (ECL) is coupled to a 1 mm long SOH MZM. The device features a π-voltage of 0.74 V at DC, measured at a bias voltage of around 3.5 V to prevent screening effects [71]. At the output of the modulator chip, the optical signal is amplified by an erbium-doped fiber amplifier (EDFA) to compensate for the insertion loss of the MZM chip. The total fiber-to-fiber insertion loss of our current devices amounts to approximately 20 dB and is dominated by the fiber-chip coupling loss of non-optimized grating couplers of about 6 dB per interface. The on-chip loss of the MZM amounts to approximately 8 dB for maximum transmission of the modulator and can be decomposed into losses of passive components (< 1 dB), free-carrier absorption within the doped silicon waveguides of 1.5 dB, and scattering loss of 5.5 dB due to sidewall roughness of the slot waveguide. These losses can be reduced considerably in future device generations. It has been shown that the roughness-induced propagation loss of strip-loaded slot waveguides can be diminished to 0.2 dB/mm by improved fabrication procedures and by using asymmetric cross sections [164]. Similarly, optimization of the doping profiles can lead to carrier-induced propagation losses of less than 1 dB/mm such that the total insertion loss of the 1 mm-long slot-waveguide section amounts to less than 1.2 dB. Each phase shifter of the MZM consists of a slot waveguide and two rib-to-slot converters, each of which contributes another 0.02 dB of insertion loss [69].

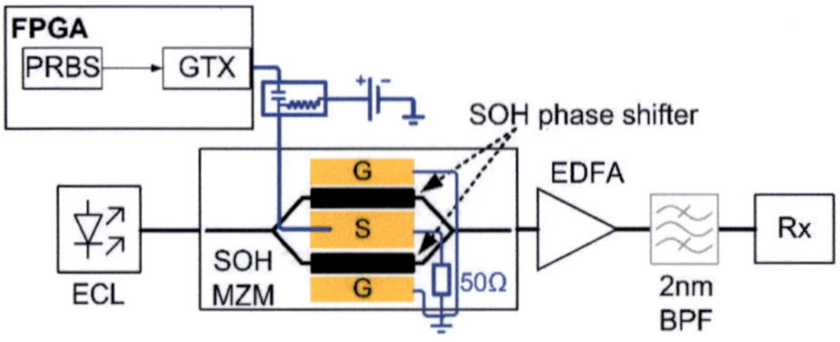

Figure 5.2: Experimental setup for directly driving an SOH modulator with the GTX interfaces of a Virtex-7 FPGA. A pseudo-random binary sequence (PRBS) at the GTX output ports is applied to the Mach-Zehnder modulator (MZM) without further amplification. The MZM consists of two 1 mm long SOH phase shifters that are driven by a ground-signal-ground (GSG) transmission line. Drive signal and bias voltage are combined in a bias-tee. The optical carrier from an external cavity laser (ECL) is modulated in the SOH MZM, and an erbium-doped fiber amplifier (EDFA) compensates the insertion loss before the signal is fed to the receiver (Rx). For coherent detection, an optical modulation analyzer (OMA) is used, whereas direct detection relies on a single photodiode. For direct detection, a 2 nm bandpass filter (BPF) suppresses ASE noise of the EDFA.

Power splitting and combining is accomplished by a pair of multi-mode interference couplers (MMI), for which the insertion losses can be as low as 0.2 dB per coupler [75]. This would amount to total on-chip insertion losses of less than 1.7 dB for an SOH MZM with 1 mm-long phase shifter sections. For packaged devices, we may assume fiber-to-chip coupling losses of less than 1.7 dB per interface, enabled by optimized grating couplers [182] or by photonic wire bonds [128], and propagation loss of less than 0.2 dB in few millimeters of low-loss access waveguides [183]. We hence expect that fiber-to-fiber insertion losses can be reduced to less than 5.3 dB for an MZM. These estimations are entirely based on already demonstrated structures; further improvements will be possible as better components become available. As a consequence, the EDFA can be omitted for data transmission using future device generations. In this context, the power handling capability of the device appears as an important parameter, which is directly linked to the photostability of the organic material. This aspect is subject to ongoing investigations. In our experiment, we operated each of the SOH phase shifters with optical powers of more than 5 dBm over several hours without observing any photo-induced deterioration of the device performance. This power was not limited by the photostability of the EO material, but by the output power of the transmitter laser and the insertion loss of the device. We expect that systematic investigations of the photostability will exhibit damage thresholds that are significantly larger. Still, the currently used 5 dBm of on-chip power per phase shifter would correspond to a total output power of more than 7.6 dBm, taking into account the insertion losses of optimized MMI power combiners and assuming full transmission of the MZM. This should enable EDFA-less transmission in future devices.

After amplification, the signal is fed to the optical receiver. The receivers are chosen differently depending on the modulation format: A coherent receiver (Rx) using another ECL as a local oscillator is used for coherent BPSK demodulation, whereas direct detection on a fast photodiode is performed for OOK signals. In the case of direct detection, a 2 nm bandpass filter (BPF) suppresses amplified spontaneous emission (ASE) from the optical amplifier.

For the experiment, we use a Xilinx XC7VX485T FPGA on a VC707 evaluation board. A pseudo-random binary sequence (PRBS) of length $2^{31} - 1$ is generated on the FPGA and fed to a single GTX transmitter (Tx) for OOK and BPSK modulation. GTX transceivers are specified to operate at a maximum of 0.5 V_{pp} for single-ended operation in the low-frequency limit. In our case, we measure a swing of about 310 mV_{pp} from the electric eye diagram at the transmitter output. The Xilinx GTX interfaces are neither designed to drive photonic devices nor do they provide a flat frequency response. The pronounced low-pass characteristic was compensated by an internal digital 3-tap finite-impulse response (FIR) filter. Using the FIR filter configured to high-pass behavior results in a nicely open eye diagram of the electrical signal, Figure 5.3 (a), but further reduces the voltage swing at the output port to 270 mV_{pp} for our experiment. For the optical signal generation experiment, the FPGA outputs are directly connected to microwave probes that contact the modulator's GSG transmission lines. A bias voltage is applied via bias-tee to adjust the operation point.

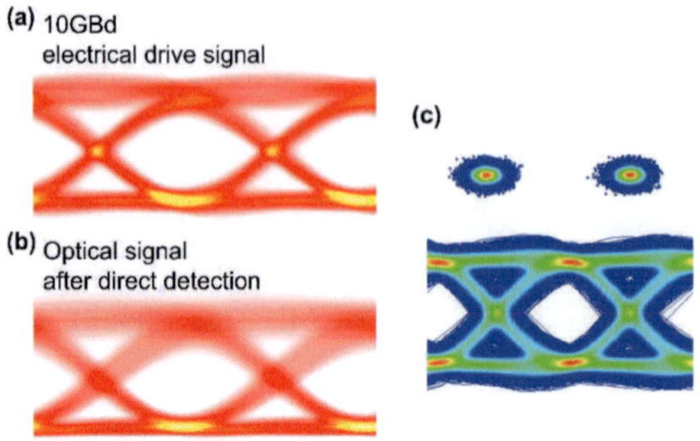

Figure 5.3: Electrical drive signals and corresponding optical signals for OOK and BPSK at 10 Gbit/s. (a) Eye diagram of electrical output of GTX transmitter. (b) Eye diagram of the optical OOK signal after direct detection by a photodiode. The quality factor Q = 6.7 indicates error-free detection. (c) Constellation diagram (top) and eye diagram (bottom) of optical BPSK signal obtained after coherent reception. An error-vector magnitude (EVM) of 13.7 % indicating error-free signaling [51] (BER < 10^{-9}) is measured.

5.3.1 Signal generation

To characterize the performance of the transmitter, OOK and BPSK signals are generated and recorded for offline signal analysis. The experimental results are summarized in Figure 5.3 (b) and (c). OOK signals are obtained by biasing the SOH MZM at the quadrature point. A real-time oscilloscope records the output current of a fast pin photodiode delivered to a 50 Ω impedance. We use Matlab for offline signal analysis. No post-equalization is employed. Eye diagrams for the electrical drive signal and the optical signal as received are depicted in Figure 5.3 (a) and (b), respectively. We do not measure any errors in approximately 390.000 recorded bits. The extracted quality-factor Q = 6.7 for the transmission indicates error-free operation (BER ≈ 10^{-11}). Comparison of both eye diagrams in Figure 5.3 (a) and (b) shows that bandwidth limitations are mainly due to the GTX module and not due to the SOH device. This is in agreement with an independent characterization of the SOH MZM, which exhibited a modulation bandwidth of approximately 15 GHz – well above the bandwidth of the GTX modules. Hence, without further bandwidth degradation, it would be possible to improve signal quality by increasing the length of the phase shifters, thereby improving modulation depth at the expense of device bandwidth. Alternatively, when keeping the device geometry, this modulator could be used at higher symbol rates if supported by the electronic driver.

With the same drive signals, but operating the modulator in the null point, BPSK signaling is achieved. We use an optical modulation analyzer (OMA) with an external ECL acting as local oscillator (LO) for coherent detection. Signal analysis includes digital polarization alignment, frequency and phase estimation, and post-equalization. As a measure for signal quality, we use the error-vector magnitude (EVM), which represents the mean deviation of the measured constellation points from their ideal position in the complex plane [51]. For our BPSK signaling experiment, we measure an EVM of 13.7 %, which can be related to an estimate BER [51] well below 10^{-9} provided that signal performance is limited by Gaussian noise. This is consistent with the bit-error analysis of the recorded data stream of approximately 620 kbit in which no bit errors were measured. The corresponding constellation and eye diagram for the BPSK signal are shown in Figure 5.3 (c). Even without post-equalization, error-free transmission is observed: The EVM is 21.5 % (estimating BER $< 10^{-10}$), and no errors were found in the recordings. These experiments show for the first time that non-resonant silicon-based MZM can be driven directly by standard FPGA outputs without further amplification.

5.3.2 Transmission experiment

In a second experiment, we demonstrate data transmission by using the FPGA board both as a date source and sink. We again transmit OOK signals and detect them by a high-speed photodiode, but now connect the photodiode output to a GTX receiver on the same board, Figure 5.4, thus simulating an electrical end-to-end communication system with optical transmission interface. No additional electrical amplifiers are used at the transmitter or receiver side. EDFA and BPF are again used for loss compensation and elimination of ASE. Signal generation and reception is performed with the Chipscope software provided by Xilinx. We measure a real-time BER of 6.6×10^{-6}.

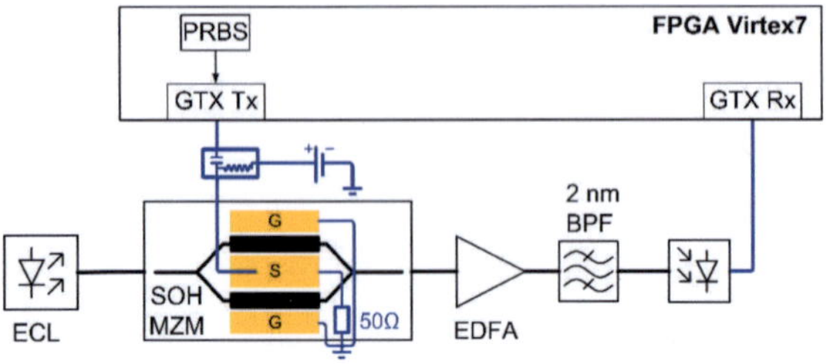

Figure 5.4: Experimental setup for OOK transmission experiment: Laser light from an ECL is coupled to the SOH modulator that is directly driven from the GTX Tx output of the FPGA. The signal is amplified by an erbium-doped fiber amplifier (EDFA), and a band-pass filter with a 2 nm passband is used to suppress amplified spontaneous emission (ASE) noise. A photodiode is used for direct detection of the signal. The electrical signal extracted from the photodiode and sent to a GTX Rx port of the same FPGA without further electrical amplification. Data analysis showed BER of 6.6×10^{-6}.

This is significantly worse than in the transmission experiments with offline data processing, but still well above typical forward-error correction (FEC) thresholds of 4.5×10^{-3} for a 7% overhead [133]. The lower reception quality is attributed to the single-ended operation of the differential GTX Rx input, which is fed by a high-speed photodiode having rather small peak-to-peak output voltages of 30 mV_{pp}. This voltage is significantly smaller than the input voltage of 150 mV that is specified for the GTX Rx. Nevertheless, this experiment shows the general viability of an electrical end-to-end communication system with an optical interface that does not require additional electrical components.

5.4 DAC-less generation of advanced modulation formats using SOH IQ modulators

In a second set of experiments, we increase the spectral efficiency of the optical signal by employing advanced modulation formats and thus increasing the number of encoded bits per symbol while maintaining the amplifier-less scheme for signal generation. While the generation of multi-level signals, for e.g. 16QAM, usually requires digital-to-analog converters (DAC), our scheme relies solely on the binary FPGA output signals and passive components. The setup is similar to the one used in the previous experiment, see Figure 5.5. The SOH MZM modulator is replaced by a 1.5 mm long IQ modulator which is connected to the outputs of an FPGA's GTH lines operating at 13 GBd. As the device is longer than the MZM modulator, a lower π-voltage of 0.53 V is measured at DC. Considering the length difference, the values for the π-voltages of both modulators are in good agreement, leading to π-voltage-length products of less than 0.8 Vmm for both devices.

For generating each of the I and Q drive signals in the 16QAM experiment, we combine two GTH lines by using a power coupler, Figure 5.5. The 6 dB power difference between the two signal components required for a bipolar 4-level signal is realized by reducing the output voltage of one output port via the FPGA software and internal circuitry.

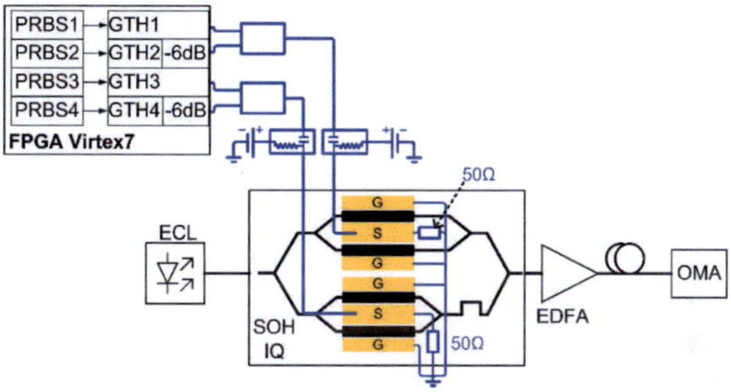

Figure 5.5: Schematic of experimental setup for DAC-less generation of 16QAM. The silicon-organic hybrid (SOH) IQ modulator consists of two 1.5 mm long Mach-Zehnder modulators (MZM). Each MZM is driven by a ground-signal-ground (GSG) transmission line contacted via a bias-tee and a microwave probe (not shown) and terminated with a 50 Ω resistor. An intentional optical path length imbalance in the parent MZM allows to adjust an optical $\pi/2$ phase shift between I and Q via wavelength tuning. To generate the four-level drive signal for each MZM, the PRBS signals from two GTH ports with independent output voltages differing by 6 dB are combined in a power combiner. The resulting 4-level signal is applied to the modulator electrodes. An external cavity laser (ECL) is used as an optical source, and an erbium-doped fiber amplifier (EDFA) compensates the optical insertion loss of the device before the signal is fed to the optical modulation analyzer (OMA).

Fine adjustment of the binary output levels of the FPGA can be used to compensate the nonlinear transfer function of the MZM. Unlike the experiment described in Section III, no FIR filter is required for compensating the low-pass characteristic of the FPGA output since the GTH output ports of the VC7222 used here exhibit a higher cut-off frequency than the GTX ports of the VC707 used in the previous experiments. Without the use of the FIR, we profit from higher output voltages. The IQ modulator features an optical insertion loss of 26 dB. Compared to the insertion loss of the MZM in Section III, the higher loss is due to longer phase-shifter sections and access waveguides. As before, device optimizations will lead to a significantly reduced insertion loss. Using the same assumptions as for the MZM, we expect that total on-chip insertion losses of less than 2.7 dB can be achieved for an IQ modulator with 1.5 mm-long slot-waveguide phase shifters. This would translate to more than 7.4 dBm of on-chip output power when limiting the optical power in the slot waveguides to 5 dBm. The EDFA can hence be omitted for data transmission. For packaged devices, fiber-to-fiber insertion losses of less than 6.3 dB are to be expected.

After amplification in a subsequent EDFA and transmission, the signal is fed to an optical modulation analyzer (OMA). The 90° phase shift between I and Q component of the optical signal is adjusted by an optical path-length imbalance between the two arms and choice of the appropriate wavelength. For simplicity, we set up a homodyne receiver by splitting the external cavity laser (ECL) output in a signal and a local oscillator (LO) path.

For the experiment, we use a Xilinx XC7VH580T FPGA on a VC7222 evaluation board. A PRBS of length $2^{31} - 1$ is generated in the FPGA and sent to the GTH ports. At a symbol rate of 13 GBd, this leads to an aggregate data rate of 26 Gbit/s and 52 Gbit/s for the QPSK and 16QAM formats, respectively. In the case of QPSK signaling, peak-to-peak drive voltages of 0.28 V_{pp} are measured for the two-level I- and Q signal before the bias-tees used to set the bias point of the modulator. For 16QAM, the peak-to-peak drive voltage of the four-level signals amounts to 0.41 V_{pp}. The bias-tees are connected to the microwave probes contacting the chip. Eye diagrams for the electrical drive signals are depicted in the top row of Figure 5.6 (a). The quality of the

16QAM drive signals is impeded by the power couplers, which are specified to work in the range of 1-18 GHz only. To quantify signal quality of the drive signal, we plot the 2- and 4-level drive signals for I and Q in the complex plane and measure the error vector magnitude of the resulting constellation diagrams that are depicted in the bottom row of Figure 5.6 (a). For measuring the EVM, we use the definition normalized to the maximum vector length of the constellation diagram denoted as EVM_m [51]. Without using an equalizer at the receiver, EVM_m values for the drive signals for QPSK and 16QAM are measured to be 13.6 % and 8.0 %, respectively. Applying an equalizer decreases the EVM_m values to 8.9 % and 6.7 %, see Figure 5.6 (a).

For the optical signal, additional to measuring EVM values, we also measure the BER. We record the signals with real-time oscilloscopes and perform offline signal analysis which includes post-equalization. Within our record length of approximately 1.6 Mbit, we do not measure any errors for the back-to-back (b2b) QPSK signal. This is in good agreement with the measured EVM_m values of 13.3 % which indicates a BER well below 10^{-9} [51]. For 16QAM b2b transmission, we measure a BER of 3.5×10^{-4} and an EVM_m of 10.2 %. The measured BER is in good agreement with the BER of 4×10^{-4} which can be estimated from the EVM [51]. The corresponding constellation diagrams are depicted on the left hand side of Figure 5.6 (b). We then perform transmission experiments with fiber spans of 10 km, 50 km and 100 km. The results of the signal quality analysis and the constellation diagrams for the b2b case and transmission are depicted in Figure 5.6 (b). An additional optical amplifier at the receiver side is used only for transmission over 100 km. For distances larger than 10 km, digital compensation of chromatic dispersion is applied. The influence of transmission on the quality of QPSK signals is negligible, as the EVM_m is still 14.3 % for a transmission distance of 100 km. No errors were found in the recordings and we estimate a BER $< 10^{-9}$ from the EVM_m. Hence, we consider the signal error-free. For the transmission of 16QAM signals, there is no significant change in EVM_m or BER for a 10 km fiber span. Even for transmission over up to 100 km, the BER of the 16QAM signal is 1.3×10^{-3}, still below the limit of 4.5×10^{-3} for 3rd generation hard-decision FEC coding with an overhead of 7 % [133].

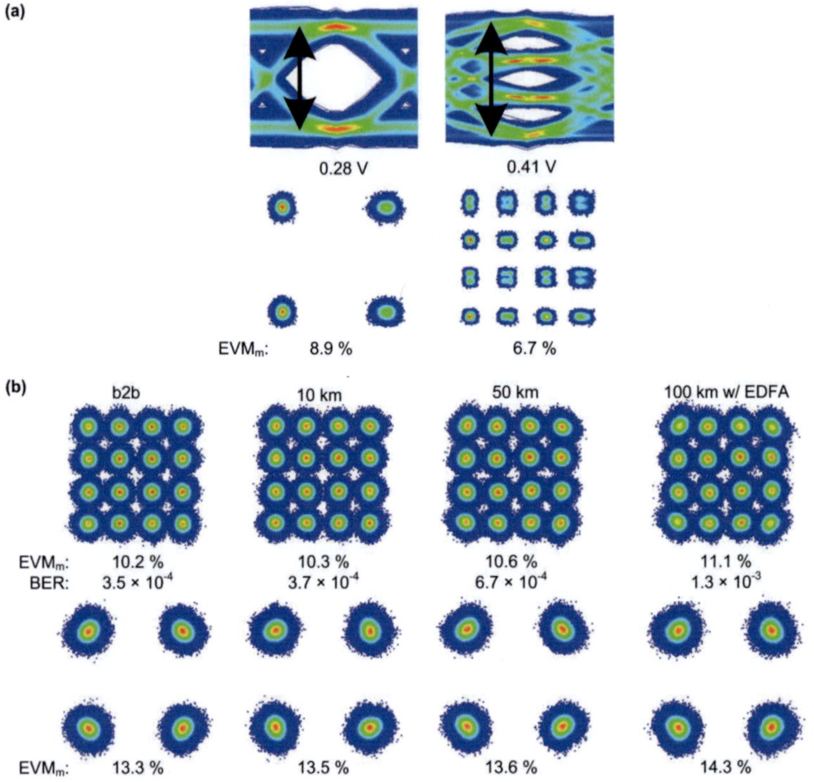

Figure 5.6: (a) Electrical drive signals from FPGA GTH ports: Eye diagrams (top) for 2- (left) and 4-level (right) drive signals for QPSK and 16QAM. Output voltages of the drive signals for QPSK and 16QAM amount to 280 mV_{pp} and 410 mV_{pp}, respectively. EVM values of the constellation diagrams (bottom) obtained after complex superposition of drive signals are measured to be 8.9 % and 6.7 % for QPSK and 16QAM, respectively after applying equalization at the receiver. (b) Optical constellation diagrams of optical 16QAM (top) and QPSK (bottom) signals generated (b2b) and transmitted over fiber spans of 10 km, 50 km and 100 km length. An erbium-doped fiber amplifier (EDFA) at the receiver side was only used for a transmission distance of 100 km. The influence of transmission distance on signal quality is small: For QPSK signals, the EVM_m of 14.3% after 100 km still suggests BER values well below 10^{-9}. For 16QAM, the measured BER increases from 3.5×10^{-4} to 1.2×10^{-3}, which is still below the 3rd generation FEC limit with 7 % overhead [133].

5.5 RF Energy consumption of modulator

In our experiments, no additional drive circuitry is used to interface the electrical ports of the FPGA to the EO modulator. As a consequence, the modulator must be operated solely with the energy provided by the binary FPGA outputs, which amounts to approximately 100 fJ/bit for a data rate of 13 Gbit/s, a peak-to-peak output voltage of 500 mV_{pp}, and a 50 Ω impedance. It is a unique feature of our SOH devices that they can be operated with a small fraction of this energy. For analysis of the modulator RF energy consumption, we assume that the terminated 50 Ω transmission lines of each MZM can be modeled as a single 50 Ω resistor in an equivalent-circuit representation. For an IQ modulator, we have to consider the power consumption of two MZMs. For 16QAM, the magnitude of the bipolar drive voltage of each MZM can have two levels that are related by a factor of 1/3. Assuming equal distribution of all 16QAM symbols, both signal levels occur with equal probability. From this, the energy consumption can be calculated by dividing the power consumption by the bit rate [119],

$$W_{\text{bit}}^{(16\text{QAM})} = 2\left[\frac{1}{2}\left(\frac{U_{\text{m}}}{2}\right)^2 \frac{1}{R_{\text{t}}} + \frac{1}{2}\left(\frac{1}{3}\frac{U_{\text{m}}}{2}\right)^2 \frac{1}{R_{\text{t}}}\right] \times \frac{1}{r} \tag{5.1}$$

where U_{m} is the peak-to-peak value of the drive voltage, $R_{\text{t}} = 50\ \Omega$ is the impedance of the termination and r is the bit rate of the 16QAM signal. Inserting the values from our experiment into the equation, $U_{\text{m}} = 0.41$ V and $r = 52$ Gbit/s, we find an energy consumption per bit of $W_{\text{bit}}^{(16\text{QAM})} = 18$ fJ/bit. This is the lowest value that has been reported for 16QAM signaling. Note that this figure solely covers the electrical energy consumption and does not take into account the power needed to drive, e.g., the laser or the EDFA in our experiment. We expect that the EDFA can be omitted completely when using optimized devices with significantly reduced insertion loss.

5.6 Summary

We report on a proof-of-concept experiment of DAC-less and amplifier-less generation and transmission of optical data signals by directly driving SOH modulators with the binary sub-Volt outputs of an FPGA. Using an SOH Mach-Zehnder modulator, we generate error-free OOK and BSPK signals at 10 Gbit/s. We demonstrate the viability for end-to-end transmission by feeding the OOK signal to a photodiode which is connected to a GTX receiver port of the same FPGA. In a second set of experiments, we generate QPSK and 16QAM signals at 13 GBd and transmit them over distances of up to 100 km. QPSK remains error-free for all transmission distances; for 16QAM, the BER remains below the threshold for hard-decision forward-error correction with 7 % overhead. The SOH IQ modulator is shown to be highly energy-efficient, enabling 16QAM signaling at an RF energy consumption of only 18 fJ/bit at a data rate of 52 Gbit/s. To the best of our knowledge, this is the lowest value reported for 16QAM signaling in literature so far.

[end of paper [J7]]

6 Summary and outlook

6.1 Summary

Photonic integration is key for the realization of compact and energy-efficient optical communication modules. Specifically, silicon photonics is a promising candidate for co-integration of electronics and photonics. This thesis discusses the suitability of silicon-organic hybrid (SOH) modulators for high-speed and low-power applications. The SOH approach combines silicon photonic slot waveguides and organic electro-optic cladding materials. The silicon structures can be fabricated in commercial lithography processes [21] and the cladding material can be deposited in a subsequent processing step. SOH modulators outperform conventional silicon photonic as well as InP modulators in terms of their modulation efficiency – quantified by the π-voltage-length product $U_\pi L$ – by a factor of approximately 20 and 10, respectively. At the same time, the devices allow for data communications at the highest symbol rates demonstrated for semiconductor-based optical modulators to date. Yet, while investigations regarding reliability and stability of SOH modulators are ongoing, this work demonstrates the suitability of the SOH concept for use in different fields of application for optical communications and achieved a number of record results.

SOH modulators for high-speed operation

Operation of SOH modulators for symbol rates of up to 100 GBd have been demonstrated for both, intensity modulation and coherent modulation formats. These are the highest symbol rates demonstrated for semiconductor-based optical modulators. The unrivalled modulation efficiency allows to operate at the lowest voltages reported for semiconductor-based modulators. Both the 100 Gbit/s OOK and the 400 Gbit/s 16QAM signal represent record demonstrations for silicon-based optical modulators.

SOH modulators for low-power applications

Besides the ability for high-speed signaling, SOH modulators are operated in an amplifier-less and DAC-less scheme: it has been shown that SOH modulators can be directly driven from the binary outputs of an FPGA with drive voltages as low as 280 mV_{pp}. Electrical amplification can be omitted for this purpose. Multi-level drive signals for the generation of 16QAM signals were realized by passive RF power combiners. For the 13 GBd 16QAM signal, a record-low electrical energy consumption of only 18 fJ/bit in the modulator is calculated.

6.2 Outlook

Concerning the integration and fabrication of SOH modulators, previous work demonstrated that SOH modulators can be fabricated in commercial lithographic fabrication processes [21] and the experimental results in this thesis and other demonstrations clearly witness the suitability of the SOH concept for future application in optical communication systems and data centers.

Yet, it is to be addressed and demonstrated that SOH modulators can withstand the environmental conditions – particularly heat – evident in typical application environments. This question particularly refers to the long-term stability of the organic materials. The material used in this work has already been demonstrated to allow for operation of SOH modulators at an operating temperature of 80 °C [22] and reported to maintain more than 90 % of its electro-optic activity for 500 hours at 85 °C [80]. However, 500 hours do not suffice to fulfill standard tests such as Telcordia. Nevertheless, the tremendous progress in material synthesis in recent years leads to the expectation that various materials combing good electro-optic stability combined with long-term stability in both, thermal relaxation and photobleaching, are to come in the near future. With successful demonstration of long-term stability, all required properties of SOH modulators for the application in commercial systems would be available. Consequently, the next steps would be device optimization, packaging and commercialization of the SOH approach.

Further, it is unclear whether the so far continued process of moving to ever higher and higher symbol rates can be sustained much longer: With the 100 GBd operation, the optical signals have reached the ability of the fasted electrical test equipment on the market to date [166]. In conjunction with interests in commercialization, next steps are rather likely to reduce the symbol rate to a moderate level and increase the spectral efficiency by using higher-order modulation formats and on-chip PDM. Building upon the efficiency of SOH modulators, the reduction of symbol rate can (and should) consequently be combined with the amplifier-less and DAC-less scheme. While in this work, the multi-level signals were generated using electrical equipment, other approaches use the optical domain; e.g., either by using segmented electrodes [16], [184] or by parallel or tandem IQ modulators [174].

Appendix

A. Pockels effect and electro-optic modulators

A.1 Pockels effect

Conventional optical phase modulators rely on a non-linear interaction, more specifically on the linear electro-optic effect, also called Pockels effect. The following discussion is based on the derivation in [60] and a more detailed discussion of optical nonlinearities can be found in [61].

The response of a bulk material to an imposing electric field $\vec{E}$ is described by the material polarization $\vec{P}$. The spatially local polarization in time domain can be described by a Volterra series which accounts for potential memory effects in the nonlinear medium [61],

$$\begin{aligned}\vec{P}(\vec{r},t) &= \varepsilon_0 \int_{-\infty}^{+\infty} \underline{\chi}^{(1)}(\vec{r},t-t')\vec{E}(r,t')\mathrm{d}t' \\ &+\varepsilon_0 \int_{-\infty}^{+\infty}\int_{-\infty}^{+\infty} \underline{\chi}^{(2)}(\vec{r},t-t',t-t''):\vec{E}(r,t')\vec{E}(r,t'')\mathrm{d}t'\mathrm{d}t'' \\ &+\varepsilon_0 \int_{-\infty}^{+\infty}\int_{-\infty}^{+\infty}\int_{-\infty}^{+\infty} \underline{\chi}^{(3)}(\vec{r},t-t',t-t'',t-t'''):\vec{E}(r,t')\vec{E}(r,t'')\vec{E}(r,t''')\mathrm{d}t'\mathrm{d}t''\mathrm{d}t''' \\ &+\ldots,\end{aligned} \tag{A.1}$$

where the vacuum permittivity is denoted by ε_0 and the response functions $\underline{\chi}^{(n)}(\vec{r},t',t'',...,t^{(n)})$ are tensors of rank $n+1$. As to the causality of the material we find $\underline{\chi}^{(n)}(\vec{r},t',t'',...,t^{(n)})=\underline{0}$ for $t',t'',t'''\ldots<0$. Eq. (A.1) can be interpreted as a linear and a non-linear interaction of matter with the electric field,

$$\begin{aligned}\vec{P}\left(\vec{r},t\right) &= \vec{P}^{(1)}\left(\vec{r},t\right)+\vec{P}^{(2)}\left(\vec{r},t\right)+\vec{P}^{(3)}\left(\vec{r},t\right)+\dots \\ &= \vec{P}^{(1)}\left(\vec{r},t\right)+\vec{P}_{\mathrm{NL}}\left(\vec{r},t\right).\end{aligned} \tag{A.2}$$

The following discussion is limited to second-order nonlinearities and particularly to the Pockels effect which describes the change of the refractive index of the material in dependence of a low frequency modulating field with angular frequency $\omega_{\mathrm{m}} \ll \omega_0, \omega_{\mathrm{m}} \approx 0$. Taking these simplifications into account, the second-order polarization term in the frequency-domain can be written as [60]

$$\tilde{P}_i^{(2)}(\omega_0+\omega_m \approx \omega_0) = 2\varepsilon_0 \sum_{jk} \tilde{\chi}_{ijk}^{(2)}(\omega_0+\omega_m;\omega_0,\omega_m)\tilde{\vec{E}}_j(\omega_0)\tilde{\vec{E}}_k(\omega_m), \tag{A.3}$$

where the indices $i,j,k=\{1,2,3\}$ denote the Cartesian components. In the above equation, the spatial dependency was omitted for improved readability.

For the electric displacement $\vec{D}$,

$$\vec{D}\left(\vec{r},t\right) = \varepsilon_0\varepsilon_r\vec{E}\left(\vec{r},t\right) = \varepsilon_0\vec{E}\left(\vec{r},t\right)+\vec{P}\left(\vec{r},t\right) \tag{A.4}$$

with the relative permittivity ε_r, and using Einstein notation for sums the electric displacement can then be expressed as [60]

$$\begin{aligned}\tilde{D}_i(\omega_0) &= \varepsilon_0\left(\delta_{ij}+\tilde{\chi}_{ij}^{(1)}(\omega_0)\right)\tilde{E}_j(\omega_0) \\ &\quad +2\varepsilon_0\tilde{\chi}_{ijk}^{(2)}(\omega_0+\omega_{\mathrm{m}};\omega_0,\omega_{\mathrm{m}})\tilde{E}_j(\omega_0)\tilde{E}_k(\omega_{\mathrm{m}}).\end{aligned} \tag{A.5}$$

Using the second-order susceptibility $\tilde{\chi}_{ijk}^{(2)}$, the change of permittivity $\Delta\varepsilon_r$ due to the Pockels effect can then be described by

$$\Delta\varepsilon_{\mathrm{r},ij} = 2\tilde{\chi}_{ijk}^{(2)}\tilde{E}_k(\omega_{\mathrm{m}}). \tag{A.6}$$

The relative permittivity is related to the refractive index of the material, $\varepsilon_{\mathrm{r}} = n^2$. For small changes, we use the approximation $\varepsilon_{\mathrm{r}}+\Delta\varepsilon_{\mathrm{r}} = (n+\Delta n)^2 \approx n+2n\Delta n$ to quantify the change of refractive index:

$$\Delta n_{ij} = \frac{\tilde{\chi}_{ijk}^{(2)}}{n} \tilde{E}_k(\omega_{\mathrm{m}}). \tag{A.7}$$

Further, the second-order susceptibility can be transformed into the commonly used electro-optic coefficient [60]

$$r_{ijk} = -\frac{2\tilde{\chi}_{ijk}^{(2)}}{n^4}. \tag{A.8}$$

that relates to the change of refractive index according to

$$\Delta n_{ij} = -\frac{1}{2} n^3 r_{ijk} \tilde{E}_k(\omega_{\mathrm{m}}). \tag{A.9}$$

The elements of the tensor of rank $n+1=3$, however, are not independent and can hence it be further simplified by taking into account the intrinsic permutation symmetry of the fields and the Kleinman symmetry [61]. Therefore, we can introduce matrices for the electro-optic coefficient r_{il} and the susceptibility $\tilde{\chi}_{il}^{(2)}$ where the indices j,k can be replaced by an index l according to

$$\begin{array}{lcccccc} jk: & 11 & 22 & 33 & 23,32 & 13,31 & 12,21 \\ l: & 1 & 2 & 3 & 4 & 5 & 6 \end{array} \tag{A.10}$$

A.2 Electro-optic modulators

The most commonly used electro-optic modulators rely on lithium niobate ($LiNbO_3$) as the nonlinear material. For practical reasons, both, the optical field and the modulating field E_{m} are typically oriented in transverse direction such that r_{33} is the relevant electro-optic coefficient. The change in refractive index Δn and the phase shift φ_{PM} in a modulator of length L can then be written as

$$\Delta n = -\frac{1}{2} n^3 r_{33} E_{\mathrm{m}}. \tag{A.11}$$

$$\varphi_{\mathrm{PM}} = \frac{1}{2} n^3 r_{33} E_{\mathrm{m}} k_0 L, \tag{A.12}$$

where $k_0 = 2\pi/\lambda_0$ is the vacuum wave number at vacuum λ_0 and the modulating electric field E_m can also be expressed by the modulating voltage U_m and the separation d of the electrodes

$$E_\mathrm{m} = \frac{U_\mathrm{m}}{d}. \tag{A.13}$$

Consequently, the phase shift arising from the Pockels effect is proportional to the applied voltage. A common figure of merit for an electro-optic phase shifter hence is the π-voltage $U_{\pi,\mathrm{PM}}$

$$U_{\pi,\mathrm{PM}} = \frac{d}{L}\frac{\lambda_0}{r_{33}n^3} \tag{A.14}$$

for which the optical phase is shifted by $\varphi_\mathrm{PM} = \pi$. Similarly, a π-voltage $U_{\pi,\mathrm{MZM}}$ can be defined for a push-pull Mach-Zehnder structure with two parallel phase shifters. It is commonly defined by a relative phase difference in of $\Delta\varphi_\mathrm{MZM} = \varphi_1 - \varphi_2 = 2\varphi = \pi$ the two arms. For an MZM we modify Eq. (A.14) accordingly:

$$U_{\pi,\mathrm{MZM}} = \frac{d}{2L}\frac{\lambda_0}{r_{33}n^3} \tag{A.15}$$

In this work, we use the term U_π for the π-voltage of a push-pull Mach-Zehnder and omit the subscript 'MZM'.

From Eqs. (A.14), (A.15) it is evident that the π-voltage can be traded for device length. Therefore we use another quantity, the π-voltage-length product $U_\pi L$, to quantify the modulation efficiency:

$$U_\pi L = \frac{d}{2}\frac{\lambda_0}{r_{33}n^3} \tag{A.16}$$

B. Digital signal processing in coherent communications

While coherent reception has been discussed already in the 1970's [185], coherent communications has not been applicable until the end of the first decade of the 21st century. This is because the compensation of any signal impairment apparent during generation, transmission, and reception of coherent optical signals was hardly to be handled by the available hardware. In contrast, with digital signal processing a chain of algorithms could be developed which only required an acceptable amount of complexity. These impairments such as chromatic dispersion and polarization changes during fiber transmission and the phase noise of the transmitter and LO laser have been neglected in the discussions above, but need to be corrected upon reception. This section – without claiming completeness – shall give an overview on aspects in digital signal processing for optical communications, and shall refer to more detailed literature for the interested reader. Compact overviews on digital signal processing are found in references [170] and [186].

A diagram of the different processing blocks is shown in Figure B.1. According to [55], on the transmitter side, binary data are divided into two parallel data streams representing the signals for the two orthogonal polarizations. The data are mapped according to the desired modulation format as discussed in Section 2.1.2 and pulse-shaping, see Section 2.1.3, is applied before the signal is converted to the analog domain and drives an optical modulator. The signal is then transmitted through the optical fiber leading to chromatic dispersion, polarization rotation, fiber nonlinearities, and others. At the output of the optical receiver, the in-phase and quadrature components of the two polarizations are converted to the digital domain using analog-to-digital converters (ADC) and subsequently pass a number of DSP subsystems.

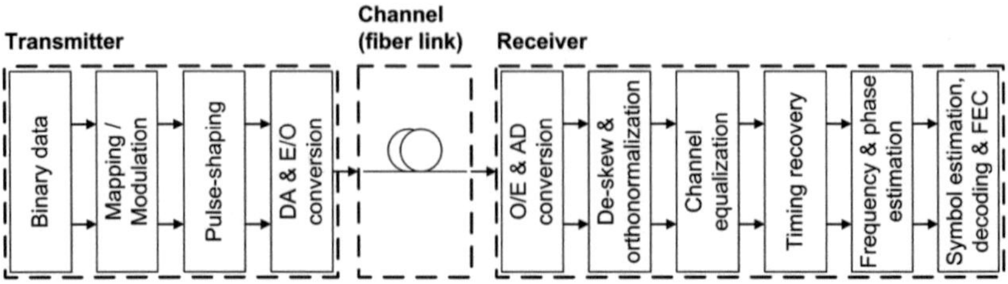

Figure B.1: Schematic overview of transmitter and receiver DSP-subsystems. A binary data stream is split in to parallel paths representing the two polarizations. Data streams are mapped according the modulation format and the pulse shape is generated before digital-to-analog (DA) and electro-to-optical (E/O) conversion. After transmission through the fiber link, the signal is received and converted to the electrical (O/E) and digital (AD) domain. De-skewing and orthonormalization account for unequal loss and propagation delays of the setup. The channel equalization compensates the frequency response of the components and physical effect on the optical link such as dispersion and polarization rotation. To recover the digital information, timing recovery and estimation of the frequency and phase differences is performed before demodulation and forward error correction (FEC) coding.

At first, different optical or electrical path lengths and propagation losses in the receiver frontend due to imperfect matching (de-skewing) as well as impairments due to deviations from the 90° phase difference of the I and Q component of the 90°-hybrid (orthonormalization) are corrected in a DSP block for *de-skewing and orthonormalization* [186].

The second DSP block is about channel equalization and may be split into another set of subsystems. References by Seb Savory, [170] and [186], discriminate into a static and dynamic channel equalization. The reasoning behind this separation is that different properties may change at different time constants and thus can be tracked at different rates: On a millisecond timescale, chromatic dispersion and the frequency response of the system may be considered constant, whereas the state of polarization may vary [170]. Other authors subdivide into even more subsystems such as dispersion compensation, polarization demultiplexing, matched filtering, and a separate equalizer at the end of the processing chain as described in [55]. Chromatic dispersion represents a linear process and can therefore be compensated using a linear digital filter. A detailed description can be found in [55], [170], [186], [187]. References [55],

[170], [186] also cover polarization demultiplexing and the compensation of polarization mode dispersion (PMD). The digital filters discussed in the references given so far cover linear effects and use linear filters for the compensation thereof. However, they are not suited to compensate for signal impairments caused by optical nonlinear interaction [186]. More recently, non-linear equalization techniques have attracted more attention. Such equalization techniques can additionally compensate for non-linear transfer characteristics as, e.g. for optical modulators and for memory effects in RF equipment and optical nonlinear interaction [188]–[190].

A timing recovery algorithm finds the correct sampling time in order to read the data signals. This is especially important for pulse-shaped signals, where a slight deviation may already lead to noticeable inter-symbol interference (ISI) [53]. Timing recovery typically requires a resampling (or interpolation) of the signal and timing information can be retrieved in data-aided and non-data-aided schemes [186]. In a next step, frequency and phase estimation is performed. First, as apparent from Eq. (2.16), the frequency offset between the optical signal an the LO needs to be compensated in order to retrieve the I and Q-data. Further, the phase term $\Delta\varphi(t)$ includes phase terms from the modulation and also contains phase noise contribution from the signal and LO lasers. Theses phase fluctuations need to be compensated in order to reliably read the data stream. Phase noise compensation is obviously dependent on the strength of phase noise, quantified by the linewidth of the Tx and Rx laser but also on the symbol rate: stronger phase noise can be tolerated for higher symbol rates. Therefore, a figure-of-merit is the product of linewidth and symbol duration which also depends on the choice of the modulation format [191]. A vast number of approaches for various modulation formats have been discussed [192]–[195] and a an overview is given in [186].

In a last step, the received symbols need to be decoded into bit streams. The decoding can be directed by a hard-decision process in which the received symbol is assigned to the closest (ideal) constellation point or by a soft-decision procedure which also includes statistical magnitudes as an input to the decision. The symbols may be received falsely, meaning that they were assigned to the wrong constellation point such that bit errors occur. To ensure error-free data transmission, coherent communication commonly relies on the use of forward-error correction (FEC) algorithms. FEC codes in general add a certain amount of overhead onto the net amount of data. The signal overhead consequently reduces the effective data rate, also referred to as net data rate. FEC codes differ in the amount of overhead that is required but also in their efficiency. The efficiency, commonly referred to as net coding gain (NCG) describes by which proportion the signal is virtually improved in its SNR and is referred to a certain output BER. In more practical terms, the NCG is a quantity that describes at which input BER the signal can be corrected to a certain output BER. A number of algorithms and generations have been discussed. Today's FEC algorithms typically handle input BERs of 4.5×10^{-3} and 2×10^{-2} with an overhead of 7 % and ~20%, respectively. More details on FEC codes can be found in the following references: [133], [168], [169], [196]–[198].

C. Quality-factor

In Section 2.1.1, the quality factor (Q-factor) was introduced as a quality metric for two-level electrical data signals. In this section, the prerequisites for the conversion to a bit error ratio (BER) will be illuminated in more detail. In principle, the BER can be calculated using the probabilities $p_{\mathrm{b}}(0\mathrm{t})$, $p_{\mathrm{b}}(1\mathrm{t})$ with which a logical '0' and '1' were transmitted and the probabilities for false detection, i.e., if either a logical '1' was transmitted (1t), but detected as a logical '0' (0d) or vice versa (0t, 1d) [199]

$$\mathrm{BER} = p_{\mathrm{b}}(1\mathrm{t})\,p_{\mathrm{b}}(0\mathrm{d}|1\mathrm{t}) + p_{\mathrm{b}}(0\mathrm{t})\,p_{\mathrm{b}}(1\mathrm{d}|0\mathrm{t}). \tag{C.1}$$

The probabilities for false detection can be expressed using the probability density functions (PDF) for the two states, w_0, w_1, and the decision threshold u_{th},

$$\mathrm{BER} = p_{\mathrm{b}}(1\mathrm{t}) \int_{-\infty}^{u_{\mathrm{th}}} w_1(u)\,du + p_{\mathrm{b}}(0\mathrm{t}) \int_{u_{\mathrm{th}}}^{\infty} w_0(u)\,du, \tag{C.2}$$

The minimum BER is found for an optimum decision threshold for which it holds $\partial \mathrm{BER}/\partial u_{\mathrm{th}} = 0$, and one finds

$$p_{\mathrm{b}}(1\mathrm{t})\,w_1(u_{\mathrm{th}}) = p_{\mathrm{b}}(0\mathrm{t})\,w_1(u_{\mathrm{th}}). \tag{C.3}$$

Consequently, for $p_{\mathrm{b}}(1\mathrm{t}) = p_{\mathrm{b}}(0\mathrm{t}) = 1/2$ the optimum decision threshold is found at the intersection of both (monomodal) PDFs.

In the following, we assume that the (mean) signal levels u_0, u_1 for the logical '0' and logical '1' are superimposed with additive Gaussian noise. The probability density functions for each of the two signal states can then be given as

$$w_0(u) = \frac{1}{\sqrt{2\pi}\sigma_0} \exp\left(\frac{(u-u_0)^2}{2\sigma_0^2} \right), \tag{C.4}$$

$$w_1(u) = \frac{1}{\sqrt{2\pi}\sigma_1} \exp\left(\frac{(u-u_1)^2}{2\sigma_1^2}\right). \tag{C.5}$$

For Gaussian PDFs, the probability of false detection can be written using the complementary error function erfc, see Eq. (2.8),

$$p_\mathrm{b}(1\mathrm{d}|0\mathrm{t}) = \int_{-\infty}^{u_\mathrm{th}} w_1(u)\,du = \frac{1}{2}\mathrm{erfc}\left(\frac{u_1 - u_\mathrm{th}}{\sigma_1\sqrt{2}}\right) \tag{C.6}$$

$$p_\mathrm{b}(0\mathrm{d}|1\mathrm{s}) = \int_{u_\mathrm{th}}^{-\infty} w_0(u)\,du = \frac{1}{2}\mathrm{erfc}\left(\frac{u_\mathrm{th} - u_0}{\sigma_0\sqrt{2}}\right). \tag{C.7}$$

Further, defining the Q-factor, [199]

$$Q = \frac{u_1 - u_0}{\sigma_0 + \sigma_1} = \frac{u_1 - u_\mathrm{th}}{\sigma_1} = \frac{u_\mathrm{th} - u_0}{\sigma_0}, \tag{C.8}$$

we can rewrite Eqs. (C.6), (C.7) and include the Q-factor. For equal probability of the two symbols, we can relate the Q-factor to a BER,

$$p_\mathrm{b}(0\mathrm{d}|1\mathrm{t}) = p_\mathrm{b}(1\mathrm{d}|0\mathrm{t}) = \frac{1}{2}\mathrm{erfc}\left(Q/\sqrt{2}\right), \tag{C.9}$$

$$\mathrm{BER} = \frac{1}{2}\mathrm{erfc}\left(\frac{Q}{\sqrt{2}}\right). \tag{C.10}$$

It shall be noted that the use of the complementary error function is a direct consequence of the Gaussian probability density functions and, consequently, Eq. (C.10) only holds for signals superimposed with (additive) Gaussian noise. It particularly holds for signals which are dominated by electrical noise. In contrast for signals dominated by optical noise, the (Gaussian) noise properties are not conserved in the power-law detection during direct detection on a photo diode. It can be shown that the distribution in the "On"-state is Gaussian-like, but for the "Off"-state of an OOK signal the distribution is exponential. In this case an analytic expression cannot be given, but the problem can only be solved numerically.

D. Characteristics of SOH modulators

D.1 Chirp Analysis

For transmission of intensity-modulated signals, an unwanted phase modulation, also referred to as chirp, may leads to increased sensitivity with respect to fiber dispersion and may hence leads to a signal quality penalty [200]. Ideally, SOH MZM should not exhibit any chirp – the field-induced refractive index change of the polymer cladding according to the Pockels effect [118] allows to completely suppress chirp in an ideal, perfectly balanced push-pull Mach-Zehnder modulator (MZM) [201]. For real-world devices with finite extinction ratio (ER), however, chirp is introduced by an imbalance of the optical amplitudes in the two MZM arms. For quantifying the chirp of a data signal generated by such a device, we use the chirp parameter α that is essentially defined by the ratio of the phase modulation to the amplitude modulation [134], [135], and can be estimated from a direct measurement of the static extinction ratio $\delta^{(\text{stat})}$, see Equations (3.7) and (3.8) in Section 3.3.2. To proof that finite static extinction ratio (ER) of the MZM is the dominant source of chirp in our devices, we compare the chirp parameter obtained from the ER to a direct measurement of α. To this end, we use another SOH modulator with a device layout similar to that of the device used for the data transmission experiments, but with better static extinction ratio of $\delta^{(\text{stat})} = 31$ dB.

For a direct measurement of the chirp parameter, we rely on the fiber response peak method [202]. Using a network analyzer, we measure the frequency response of the modulator with a 75 km fiber span. Erbium-doped fiber amplifiers (EDFA) are used to compensate the insertion loss of the modulator and the fiber span. The measured frequency response normalized to the back-to-back transmission is shown in Figure D.1 (a). For data evaluation, the product of the square of the resonance frequency f_μ and the fiber length L is plotted in dependence of the resonance order μ, see Figure D.1 (b). The chirp parameter α and the fiber dispersion D can then be extracted from the slope and from the vertical offset of a straight line fitted to the data points [202].

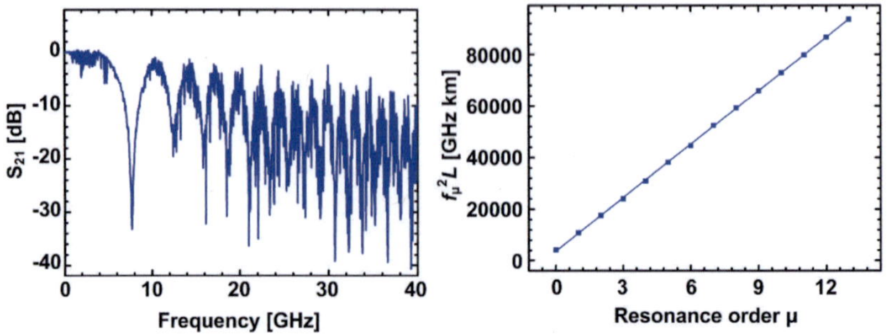

Figure D.1: Chirp characterization of SOH Mach-Zehnder modulator. (a) Frequency response of SOH MZM with 75 km fiber span for modulator operated without gate field. The frequency response has been normalized to the modulator's back-to-back characterization. Resonant dips originate from fiber dispersion. (b) From the product of the squared resonance frequency and the fiber length, $f_\mu^2 L$, fiber dispersion parameter and chirp factor α can be estimated [202]. In the analysis, we find a chirp factor of $|\alpha| \approx 0.09$.

Measurement analysis leads to a dispersion coefficient of 16.8 ps/(nm km) for the 75 km fiber span and a chirp factor α of 0.09. The measured value compares well to the value $|\alpha| = 0.06$ which we estimate directly from the measured (static) extinction ratio. This confirms that the chirp of SOH MZM is dominated by the influence of a finite ER – the slight differences are attributed to the finite measurement accuracies of the fiber response peak method.

Using the equations described above, we estimate a chirp factor $|\alpha| \approx 0.42$ for a static extinction ratio of 14 dB as obtained for the MZM used for the 100 Gbit/s data experiment in Section 3.

D.2 Wavelength operating range of SOH modulators

SOH modulators can operate over a wide range of wavelengths. In essence, the operating range is only limited by the ability of the silicon photonic slot waveguide to efficiently guide the light and by the absorption of the organic EO material. The modulator structure as well as the organic material is well suited to operate over the entire range of infrared telecommunication wavelengths, comprising all relevant transmission bands between 1260 nm and 1675 nm. Absorption of the EO material is negligible above 1200 nm [77], [79] and hence the material is transparent in wavelength regions beyond 1260 nm which are relevant for telecommunications. The performance of the slot waveguides can be quantified by considering the field interaction factor according to Eq. (3.2) in Section 3.2.1. For the device used in the transmission experiments in Section 3.3, the field interaction factor decreases from $\Gamma \approx 0.16$ at $\lambda = 1550\,\text{nm}$ to $\Gamma \approx 0.14$ at $\lambda = 1310\,\text{nm}$, see Figure D.2 for the associated mode field profiles. At the same time, the material-related figure-of-merit for EO activity, $n_{EO} r_{33}$, increases [160] by about 40 %, which leads to an overall improvement of the modulation efficiency by about 16 %. In total, the device performance does hence not change significantly with wavelength.

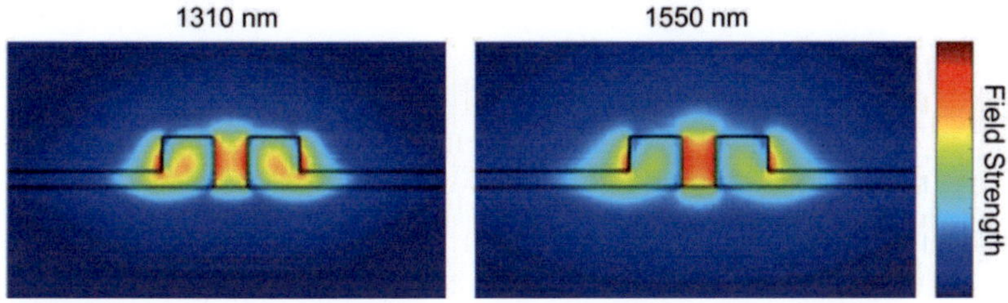

Figure D.2: Simulated mode fields for a center wavelength of 1310 nm (left) and 1550 nm (right). The color coding indicates the modulus of the transverse electrical field and is identical for both plots. For the shorter wavelength, the field is stronger in the silicon rails, and hence the field interaction factor is smaller by 17 %. This is overcompensated by an increase of the material-related figure-of-merit for EO activity, $n_{EO} r_{33}$.

E. Implementation of pulse shape for system simulation

The hardware used in Section 3.3.3 features finite rise and fall times, see the eye diagram in Figure E.1 (a) for the measured electrical output signal. For an emulation of these signals in the simulation, we use cosine-shaped pulses in the time-domain (not to be confused with pulse shapes with a raised-cosine shaped spectrum as discussed in Section 2.1.3). A mathematical description of the pulse shape $p(t)$ is given as

$$p(t)=\frac{1}{2}\left(1+\cos\left(\pi\frac{t}{T_{\mathrm{S}}}\right)\right), \qquad -T_{\mathrm{S}}\leq t\leq T_{\mathrm{S}}, \tag{E.1}$$

Where T_{s} denotes the symbol duration. With the discrete symbols $a_k=\{-1;1\}$ which correspond to a logical '1' and a logical '0', respectively, with the Dirac function $\delta(t)$ representing the time-discrete nature of digital data, and with the peak-to-peak output voltage U_{pp}, the (electrical) mean-free time-domain drive signal can be written as

$$s(t)=U_{\mathrm{pp}}\sum_{k}a_k\delta(t-kT)*p(t) \tag{E.2}$$

where $*$ denotes a convolution operation. The eye diagram of the digitally generated waveform (with additive white Gaussian noise) is depicted in Figure E.1 (b).

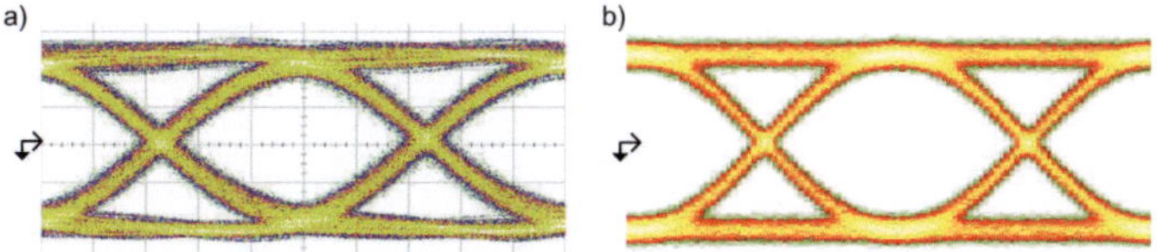

Figure E.1: Waveform for bandwidth analysis. (a) Measured eye diagram from the multiplexer (MUX) at 100 Gbit/s, having a peak-to-peak voltage swing of approximately 0.4 V_{pp}. The signal is fed to an RF amplifier and to the SOH modulator. (b) Eye diagram of digital waveform that is used to approximate the output signal of the MUX for the bandwidth simulation in Section 3.3.3.

F. Description of equipment

This section gives a detailed overview on the equipment used in the two experiments, which are referred to as Experiment 1 and Experiment 2 in Section 4.3 and 4.4.

Experiment 1:

The signal is generated by an arbitrary-waveform generator (AWG, Keysight M8196A), featuring a nominal resolution of 8 bit, an analog bandwidth of 32 GHz, and a sampling rate of up to 92 GSa/s. The data pattern is derived from a pseudo-random binary sequence (PRBS) with a length of $2^9 - 1$. We use a pulse shape with raised-cosine (RC) spectrum and roll-off factor $\beta = 0.35$. We generate QPSK signals at symbol rates up to 77 GBd and 16QAM signals up to 63 GBd. For signal generation up to 63 GBd, we use 30 GHz RF amplifiers (SHF 807) and measure an electrical eye opening of 1.0 V, which operates the modulator in its linear regime. For the 77 GBd signal, we replace the 30 GHz RF amplifiers with 50 GHz amplifiers (SHF S807) and operate at an electrical eye opening of 1.4 V. The frequency response of the RF equipment (amplifiers & cabling) is determined in a reference measurement and digitally compensated by pre-distorting the drive signal at the transmitter. At the receiver, we use a 2 nm optical bandpass filter, an optical modulation analyzer (OMA, Keysight N4391) and 63 GHz real-time oscilloscopes (Keysight DSOZ634A), each having a sampling rate of 160 GSa/s.

Experiment 2:

In Experiment 2, we use a Micram DAC4 as a signal source. The device features a nominal resolution of 6 bit, a typical analog 3 dB bandwidth of 40 GHz with a smooth roll-off to 50 GHz[166], and a sampling rate of up to 100 GSa/s. The data pattern is derived from a pseudo-random pattern with length 2^{15}. In the experiment, we did not yet use any pre-compensation of the transmitter drive signals. For 80 GBd and 100 GBd, the signal is generated without oversampling (1 sample per symbol), and the sampling rate is adjusted accordingly. Eye diagrams of electrical four-level signals at 80 GBd and

100 GBd are depicted in Figure F.1; more details can be found in [166], [203]. The 50 GBd signal is generated with two-fold oversampling (2 samples per symbol). To drive the SOH modulator, we use RF amplifiers with 70 GHz bandwidth (SHF 827) and a peak-to-peak output voltage swing of 1.5 V_{pp}. These amplifiers are actually the main source of drive signal distortion, featuring a phase response which changes by approximately 200° between 40 GHz and 75 GHz, see [121], [161] for details. In our experiment, these distortions were corrected by the receiver equalizer.. At the receiver side, we use a 1.5 nm-wide optical band pass filter, a coherent receiver (Tektronix OM4245), and 70 GHz real-time oscilloscopes (Tektronix DPO77002SX), featuring a sampling rate of 200 GSa/s.

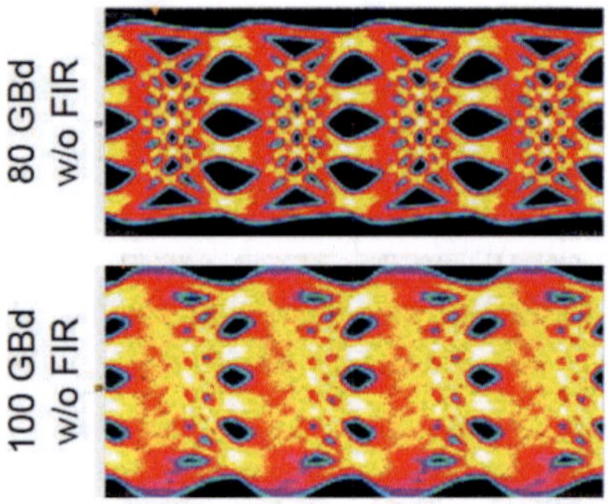

Figure F.1: Eye diagrams of electrical four-level signals at 80 GBd and 100 GBd, measured at the output of the DAC used for Experiment 2 without pre-compensation by a finite impulse response (FIR) filter.

G. Bibliography

[1] R. Ramaswami, K. Sivarajan, and G. Sasaki, *Optical Networks: A Practical Perspective*, 3rd ed. Morgan Kaufmann, 2009.

[2] C. H. Townes, "1964 Nobel Lecture: Production of Coherent Radiation by Atoms and Molecules," *IEEE Spectr.*, vol. 2, no. 8, pp. 30–43, Aug. 1965.

[3] C. K. Kao, "Nobel Lecture: Sand from Centuries Past: Send Future Voices Fast," *Rev. Mod. Phys.*, vol. 82, no. 3, pp. 2299–2303, Aug. 2010.

[4] Cisco, "Cisco Visual Networking Index: Forecast and Methodology, 2015–2020." 2016.

[5] R. C. Alferness, "Waveguide Electrooptic Modulators," *IEEE Trans. Microw. Theory Tech.*, vol. 30, no. 8, pp. 1121–1137, Aug. 1982.

[6] O. E. DeLange, "Wide-Band Optical Communication Systems: Part II - Frequency-Division Multiplexing," *Proc. IEEE*, vol. 58, no. 10, pp. 1683–1690, 1970.

[7] T. Li, "The Impact of Optical Amplifiers on Long-Distance Lightwave Telecommunications," *Proc. IEEE*, vol. 81, no. 11, pp. 1568–1579, 1993.

[8] E. L. Wooten *et al.*, "A Review of Lithium Niobate Modulators for Fiber-Optic Communications Systems," *IEEE J. Sel. Top. Quantum Electron.*, vol. 6, no. 1, pp. 69–82, Jan. 2000.

[9] M. Smit, J. van der Tol, and M. Hill, "Moore's Law in Photonics," *Laser Photon. Rev.*, vol. 6, no. 1, pp. 1–13, Jan. 2012.

[10] M. Smit *et al.*, "An Introduction to InP-Based Generic Integration Technology," *Semicond. Sci. Technol.*, vol. 29, no. 8, p. 83001, Jun. 2014.

[11] M. Hochberg *et al.*, "Silicon Photonics: The Next Fabless Semiconductor Industry," *IEEE Solid-State Circuits Mag.*, vol. 5, no. 1, pp. 48–58, 2013.

[12] S. Assefa *et al.*, "CMOS-Integrated 40GHz Germanium Waveguide Photodetector for On-Chip Optical Interconnects," in *Optical Fiber Communication Conference (OFC)*, 2009, p. OMR4.

[13] G. T. Reed, G. Mashanovich, F. Y. Gardes, and D. J. Thomson, "Silicon Optical Modulators," *Nat. Photonics*, vol. 4, pp. 518–526, Aug. 2010.

[14] J. Geyer *et al.*, "Practical Implementation of Higher Order Modulation Beyond 16-QAM," in *Optical Fiber Communication Conference (OFC)*, 2015, p. Th1B.1.

[15] P. Dong, X. Liu, S. Chandrasekhar, L. L. Buhl, R. Aroca, and Y. Chen, "Monolithic Silicon Photonic Integrated Circuits for Compact 100+Gb/s Coherent Optical Receivers and Transmitters," *IEEE J. Sel. Top. Quantum Electron.*, vol. 20, no. 4, p. 6100108, 2014.

[16] A. Shastri *et al.*, "Ultra-Low-Power Single-Polarization QAM-16 Generation Without DAC Using a CMOS Photonics Based Segmented Modulator," *J. Light. Technol.*, vol. 33, no. 6, pp. 1255–1260, Mar. 2015.

[17] C. Koos *et al.*, "Silicon-Organic Hybrid (SOH) and Plasmonic-Organic Hybrid (POH) Integration," *J. Light. Technol.*, vol. 34, no. 2, pp. 256–268, Jan. 2016.

[18] T. Baehr-Jones *et al.*, "Optical Modulation and Detection in Slotted Silicon Waveguides," *Opt. Express*, vol. 13, no. 14, p. 5216, 2005.

[19] L. Alloatti *et al.*, "100 GHz Silicon–organic Hybrid Modulator," *Light Sci. Appl.*, vol. 3, no. 5, p. e173, 2014.

[20] S. Koeber *et al.*, "Femtojoule Electro-Optic Modulation Using a Silicon–organic Hybrid Device," *Light Sci. Appl.*, vol. 4, no. 2, p. e255, Feb. 2015.

[21] M. Lauermann, "Silicon-Organic Hybrid Devices for High-Speed Electro-Optic Signal Processing," Karlsruhe Institute of Technology, 2016.

[22] M. Lauermann *et al.*, "Generation of 64 GBd 4ASK Signals Using a Silicon-Organic Hybrid Modulator at 80°C," *Opt. Express*, vol. 24, no. 9, pp. 9389–9396, May 2016.

[23] E. Snitzer, "Cylindrical Dielectric Waveguide Modes," *J. Opt. Soc. Am.*, vol. 51, no. 5, pp. 491–498, 1961.

[24] J. Zubia and J. Arrue, "Plastic Optical Fibers: An Introduction to Their Technological Processes and Applications," *Opt. Fiber Technol.*, vol. 7, no. 2, pp. 101–140, Apr. 2001.

[25] G. Parca, A. Shahpari, V. Carrozzo, G. M. T. Beleffi, and A. L. J. Teixeira, "Optical Wireless Transmission at 1.6-Tbit/s (16×100 Gbit/s) for next-Generation Convergent Urban Infrastructures," *Opt. Eng.*, vol. 52, no. 11, p. 116102, Nov. 2013.

[26] C.-Y. Lin, Y.-P. Lin, H.-H. Lu, C.-Y. Chen, T.-W. Jhang, and M.-C. Chen, "Optical Free-Space Wavelength-Division-Multiplexing Transport System," *Opt. Lett.*, vol. 39, no. 2, p. 315, Jan. 2014.

[27] E. Ciaramella *et al.*, "1.28 Terabit/s (32x40 Gbit/s) Wdm Transmission System for Free Space Optical Communications," *IEEE J. Sel. Areas Commun.*, vol. 27, no. 9, pp. 1639–1645, Dec. 2009.

[28] T. Hayashi, T. Taru, O. Shimakawa, T. Sasaki, and E. Sasaoka, "Design and Fabrication of Ultra-Low Crosstalk and Low-Loss Multi-Core Fiber," *Opt. Express*, vol. 19, no. 17, p. 16576, Aug. 2011.

[29] F. Parmigiani *et al.*, "Ultra-Flat SPM-Broadened Spectra in a Highly Nonlinear Fiber Using Parabolic Pulses Formed in a Fiber Bragg Grating," *Opt. Express*, vol. 14, no. 17, p. 7617, 2006.

[30] P. S. J. Russell, "Photonic-Crystal Fibers," *J. Light. Technol.*, vol. 24, no. 12, pp. 4729–4749, Dec. 2006.

[31] K. Kikuchi and M. Osaki, "Highly-Sensitive Coherent Optical Detection of M-Ary Frequency-Shift Keying Signal," *Opt. Express*, vol. 19, no. 26, p. B32, Dec. 2011.

[32] K.-P. Ho, *Phase-Modulated Optical Communication Systems*. New York: Springer Science+Business Media, Inc, 2005.

[33] M. Karlsson and E. Agrell, "Multilevel Pulse-Position Modulation for Optical Power-Efficient Communication Abstract :," vol. 19, no. 26, pp. 799–804, 2011.

[34] A. Ludwig *et al.*, "Stacked Modulation Formats Enabling Highest-Sensitivity Optical Free-Space Links," *Opt. Express*, vol. 23, no. 17, p. 21942, Aug. 2015.

[35] P. J. Winzer and R.-J. Essiambre, "Advanced Optical Modulation Formats," in *Optical Fiber Telecommunications V. B: Systems and Networks*, Elsevier, 2008.

[36] W. Freude *et al.*, "Quality Metrics for Optical Signals: Eye Diagram, Q-Factor, OSNR, EVM and BER," in *International Conference on Transparent Optical Networks (ICTON)*, 2012, p. Mo.B1.5.

[37] H. Yamazaki *et al.*, "160-Gbps Nyquist PAM4 Transmitter Using a Digital-Preprocessed Analog-Multiplexed DAC," *Eur. Conf. Opt. Commun.*, 2015, p. 1013.

[38] N. Stojanovic, C. Prodaniuc, F. Karinou, and Z. Qiang, "56-Gbit/s 4-D PAM-4 TCM Transmission Evaluation for 400-G Data Center Applications," *Opt. Fiber Commun. Conf.*, 2016, p. Th1G.6.

[39] J. Wei *et al.*, "Demonstration of the First Real-Time End-to-End 40-Gb/s PAM-4 for Next-Generation Access Applications Using 10-Gb/s Transmitter," *J. Light. Technol.*, vol. 34, no. 7, pp. 1628–1635, 2016.

[40] N. Eiselt *et al.*, "First Real-Time 400G PAM-4 Demonstration for Inter-Data Center Transmission over 100 Km of SSMF at 1550 Nm," *Opt. Fiber Commun. Conf.*, 2016, p. W1K.5.

[41] B. Teipen, N. Eiselt, A. Dochhan, H. Griesser, M. Eiselt, and J. P. Elbers, "Investigation of PAM-4 for Extending Reach in Data Center Interconnect Applications," in *International Conference on Transparent Optical Networks (ICTON)*, 2015, p. MO.D3.3.

[42] P. J. Winzer and R.-J. Essiambre, "Advanced Optical Modulation Formats," *Proc. IEEE*, vol. 94, no. 5, pp. 952–985, 2006.

[43] P. J. Winzer and R.-J. Essiambre, "Advanced Modulation Formats for High-Capacity Optical Transport Networks," *J. Light. Technol.*, vol. 24, no. 12, pp. 4711–4728, Dec. 2006.

[44] Infinera, "FlexCoherent® Modulation for the Terabit Age," *White Paper*. 2016.

[45] K. Schuh *et al.*, "Single Carrier 1.2 Tbit/s Transmission over 300 Km with PM-64 QAM at 100 GBaud," in *Optical Fiber Communication Conference (OFC)*, 2017, p. Th5B.5.

[46] R. Schmogrow *et al.*, "512QAM Nyquist Sinc-Pulse Transmission at 54 Gbit/s in an Optical Bandwidth of 3 GHz," *Opt. Express*, vol. 20, no. 6, pp. 6439–47, 2012.

[47] R. Maher, K. Croussore, M. Lauermann, R. Going, X. Xu, and J. Rahn, "Constellation Shaped 66 GBd DP-1024QAM Transceiver with 400 Km Transmission over Standard SMF," in *European Conference on Optical Communication (ECOC)*, 2017, p. Th.PDP.B.2.

[48] R.-J. Essiambre, G. Kramer, P. J. Winzer, G. J. Foschini, and B. Goebel, "Capacity Limits of Optical Fiber Networks," *J. Light. Technol.*, vol. 28, no. 4, pp. 662–701, 2010.

[49] K. Kikuchi, "Coherent Optical Communication Systems," in *Optical Fiber Telecommunications V. B: Systems and Networks*, Elsevier, 2008.

[50] J. Li, *Optical Delay Interferometers and Their Application for Self-Coherent Detection*. Karlsruhe: KIT Scientific Publishing, 2013.

[51] R. Schmogrow *et al.*, "Error Vector Magnitude as a Performance Measure for Advanced Modulation Formats," *IEEE Photonics Technol. Lett.*, vol. 24, no. 1, pp. 61–63, Jan. 2012.

[52] R. Schmogrow *et al.*, "Corrections to 'Error Vector Magnitude as a Performance Measure for Advanced Modulation Formats,'" *Photonics Technol. Lett. IEEE*, vol. 24, no. 23, p. 2198, 2012.

[53] J. Proakis and M. Salehi, *Digital Communications*, Fifth Edti. New York: McGraw-Hill Higher Education, 2008.

[54] N. S. Alagha and P. Kabal, "Generalized Raised-Cosine Filters," *IEEE Trans. Commun.*, vol. 47, no. 7, pp. 989–997, Jul. 1999.

[55] R. Schmogrow, *Real-Time Digital Signal Processing for Software-Defined Optical Transmitters and Receivers*. KIT Scientific Publishing, 2014.

[56] R. Schmogrow *et al.*, "Real-Time Nyquist Pulse Generation beyond 100 Gbit/s and Its Relation to OFDM," *Opt. Express*, vol. 20, no. 1, pp. 317–337, 2012.

[57] Q. Xu, B. Schmidt, S. Pradhan, and M. Lipson, "Micrometre-Scale Silicon Electro-Optic Modulator," *Nature*, vol. 435, pp. 325–327, May 2005.

[58] A. M. Gutierrez *et al.*, "Ring-Assisted Mach-Zehnder Interferometer Silicon Modulator for Enhanced Performance," *J. Light. Technol.*, vol. 30, no. 1, pp. 9–14, Jan. 2012.

[59] J. Liu *et al.*, "Waveguide-Integrated, Ultralow-Energy GeSi Electro-Absorption Modulators," *Nat. Photonics*, vol. 2, no. 7, pp. 433–437, 2008.

[60] R. Palmer, *Silicon Photonic Modulators for Low-Power Applications*, vol. 18. Karlsruhe: KIT Scientific Publishing, 2015.

[61] R. W. Boyd, *Nonlinear Optics*, 2nd ed. Academic Press, 2003.

[62] J. S. Orcutt *et al.*, "Open Foundry Platform for High-Performance Electronic-Photonic Integration.," *Opt. Express*, vol. 20, no. 11, pp. 12222–12232, May 2012.

[63] R. A. Soref and B. R. Bennett, "Electrooptical Effects in Silicon," IEEE J. Quantum Electron., vol. 23, no. 1, pp. 123–129, Jan. 1987.

[64] W. M. J. Green, M. J. Rooks, L. Sekaric, and Y. A. Vlasov, "Ultra-Compact, Low RF Power, 10 Gb/s Silicon Mach-Zehnder Modulator," *Opt. Express*, vol. 15, no. 25, p. 17106, 2007.

[65] T. Baba *et al.*, "25-Gbps Operation of Silicon P-I-N Mach-Zehnder Optical Modulator with 100-Mm-Long Phase Shifter," in *Conference on Lasers and Electro-Optics (CLEO)*, 2012, p. CF2L.3.

[66] B. Milivojevic *et al.*, "112Gb/s DP-QPSK Transmission Over 2427km SSMF Using Small-Size Silicon Photonic IQ Modulator and Low-Power CMOS Driver," *Opt. Fiber Commun. Conf.*, p. OTh1D.1, 2013.

[67] D. Petousi *et al.*, "High-Speed Monolithically Integrated Silicon Photonic Transmitters in 0.25 Mm BiCMOS Platform," in *European Conference on Optical Communication (ECOC)*, 2016, pp. 604–606.

[68] M. R. Watts, W. A. Zortman, S. Member, D. C. Trotter, R. W. Young, and A. L. Lentine, "Low-Voltage, Compact, Depletion-Mode, Silicon Mach–ZehnderModulator," vol. 16, no. 1, pp. 159–164, 2010.

[69] R. Palmer *et al.*, "Low-Loss Silicon Strip-to-Slot Mode Converters," *IEEE Photonics J.*, vol. 5, no. 1, p. 2200409, Feb. 2013.

[70] V. R. Almeida, Q. Xu, C. A. Barrios, and M. Lipson, "Guiding and Confining Light in Void Nanostructure," *Opt. Lett.*, vol. 29, no. 11, p. 1209, 2004.

[71] R. Palmer *et al.*, "High-Speed, Low Drive-Voltage Silicon-Organic Hybrid Modulator Based on a Binary-Chromophore Electro-Optic Material," *J. Light. Technol.*, vol. 32, no. 16, pp. 2726–2734, 2014.

[72] C. Kieninger *et al.*, "Record-High In-Device Electro-Optic Coefficient of 359 pm/V in a Silicon-Organic Hybrid (SOH) Modulator," in *Conference on Lasers and Electro-Optics (CLEO)*, 2017, p. STu3N.2.

[73] L. R. Dalton, P. A. Sullivan, and D. H. Bale, "Electric Field Poled Organic Electro-Optic Materials: State of the Art and Future Prospects," *Chem. Rev.*, vol. 110, no. 1, pp. 25–55, Jan. 2010.

[74] J. J. Wolff and R. Wortmann, "Organic Materials for Second-Order Non-Linear Optics," in *Advances in Physical Organic Chemistry*, vol. 32, Elsevier, 1999, pp. 121–217.

[75] R. Palmer *et al.*, "Silicon-Organic Hybrid MZI Modulator Generating OOK, BPSK and 8-ASK Signals for Up to 84 Gbit/s," *IEEE Photonics J.*, vol. 5, no. 2, p. 6600907, Apr. 2013.

[76] S. Wolf *et al.*, "DAC-Less Amplifier-Less Generation and Transmission of QAM Signals Using Sub-Volt Silicon-Organic Hybrid Modulators," *J. Light. Technol.*, vol. 33, no. 7, pp. 1425–1432, 2015.

[77] W. Heni *et al.*, "Nonlinearities of Organic Electro-Optic Materials in Nanoscale Slots and Implications for the Optimum Modulator Design," *Opt. Express*, vol. 25, no. 3, pp. 2627–2653, Feb. 2017.

[78] Y. Enami, J. Luo, and A. K. Y. Jen, "Short Hybrid Polymer/sol-Gel Silica Waveguide Switches with High in-Device Electro-Optic Coefficient Based on Photostable Chromophore," *AIP Adv.*, vol. 1, p. 42137, 2011.

[79] Y. Jouane, Y.-C. Chang, D. Zhang, J. Luo, A. K.-Y. Jen, and Y. Enami, "Unprecedented Highest Electro-Optic Coefficient of 226 pm/V for Electro-Optic polymer/TiO2 Multilayer Slot Waveguide Modulators," *Opt. Express*, vol. 22, no. 22, pp. 27725–27732, 2014.

[80] S. Huang, J. Luo, Z. Jin, X.-H. Zhou, Z. Shi, and A. K.-Y. Jen, "Enhanced Temporal Stability of a Highly Efficient Guest–host Electro-Optic Polymer through a Barrier Layer Assisted Poling Process," *J. Mater. Chem.*, vol. 22, no. 38, pp. 20353–20357, 2012.

[81] J. Luo, X.-H. Zhou, and A. K.-Y. Jen, "Rational Molecular Design and Supramolecular Assembly of Highly Efficient Organic Electro-Optic Materials," *J. Mater. Chem.*, vol. 19, p. 7410, 2009.

[82] Y. J. Cheng *et al.*, "Donor - Acceptor Thiolated Polyenic Chromophores Exhibiting Large Optical Nonlinearity and Excellent Photostability," *Chem. Mater.*, vol. 20, no. 15, pp. 5047–5054, 2008.

[83] Z. Shi *et al.*, "Achieving Excellent Electro-Optic Activity and Thermal Stability in Poled Polymers through an Expeditious Crosslinking Process," *J. Mater. Chem. C*, vol. 22, no. 3, pp. 951–959, 2012.

[84] F. Qiu and S. Yokoyama, "Efficiently Poled Electro-Optic Polymer Modulators," *Opt. Express*, vol. 24, no. 17, pp. 19020–19025, Aug. 2016.

[85] G. Gupta, W. H. Steier, Y. Liao, J. Luo, L. R. Dalton, and A. K.-Y. Jen, "Modeling Photobleaching of Optical Chromophores: Light-Intensity Effects in Precise Trimming of Integrated Polymer Devices," *J. Phys. Chem. C*, vol. 112, no. 21, pp. 8051–8060, 2008.

[86] C. Haffner *et al.*, "Harnessing Nonlinearities near Material Absorption Resonances for Reducing Losses in Plasmonic Modulators," *Opt. Mater. Express*, vol. 7, no. 7, pp. 2168–2181, 2017.

[87] L. R. Dalton and S. Benight, "Theory-Guided Design of Organic Electro-Optic Materials and Devices," *Polymers (Basel).*, vol. 3, no. 4, pp. 1325–1351, Aug. 2011.

[88] M. Klinger, L. P. Tolbod, K. V. Gothelf, and P. R. Ogilby, "Effect of Polymer Cross-Links on Oxygen Diffusion in Glassy PMMA Films," *ACS Appl. Mater. Interfaces*, vol. 1, no. 3, pp. 661–667, Mar. 2009.

[89] R. Dinu *et al.*, "Environmental Stress Testing of Electro-Optic Polymer Modulators," *J. Light. Technol.*, vol. 27, no. 11, pp. 1527–1532, Jun. 2009.

[90] R. Palmer *et al.*, "Low Power Mach-Zehnder Modulator in Silicon-Organic Hybrid Technology," *IEEE Photonics Technol. Lett.*, vol. 25, no. 13, pp. 1226–1229, Jul. 2013.

[91] L. Alloatti, C. Koos, and J. Leuthold, "Optical Loss by Surface Transfer Doping in Silicon Waveguides," *Appl. Phys. Lett.*, vol. 107, no. 3, p. 31107, Jul. 2015.

[92] L. Alloatti, M. Lauermann, C. Sürgers, C. Koos, W. Freude, and J. Leuthold, "Optical Absorption in Silicon Layers in the Presence of Charge Inversion/accumulation or Ion Implantation," *Appl. Phys. Lett.*, vol. 103, p. 51104, 2013.

[93] L. Alloatti, *High-Speed , Low-Power and Mid-IR Silicon Photonics Applications*, KIT Scientific Publishing, vol. 12. 2013.

[94] L. Alloatti *et al.*, "42.7 Gbit/s Electro-Optic Modulator in Silicon Technology," *Opt. Express*, vol. 19, no. 12, pp. 11841–11851, 2011.

[95] A. Melikyan, "Active and Passive Plasmonic Devices," Karlsruhe Institute of Technology (KIT), 2014.

[96] A. Melikyan *et al.*, "High-Speed Plasmonic Phase Modulators," *Nat. Photonics*, vol. 8, no. 3, pp. 229–233, Feb. 2014.

[97] C. Haffner *et al.*, "All-Plasmonic Mach–Zehnder Modulator Enabling Optical High-Speed Communication at the Microscale," *Nat. Photonics*, vol. 9, no. 8, pp. 525–528, 2015.

[98] C. Haffner *et al.*, "Plasmonic Organic Hybrid Modulators—Scaling Highest Speed Photonics to the Microscale," *Proc. IEEE*, vol. 104, no. 12, pp. 2362-2379, 2016.

[99] W. Heni *et al.*, "Silicon–Organic and Plasmonic–Organic Hybrid Photonics," *ACS Photonics*, vol. 4, no. 7, pp. 1576–1590, Jul. 2017.

[100] S. A. Maier, *Plasmonics: Fundamentals and Applications*. Boston, MA: Springer US, 2007.

[101] A. Melikyan *et al.*, "Plasmonic-Organic Hybrid (POH) Modulators for OOK and BPSK Signaling at 40 Gbit/s," *Opt. Express*, vol. 23, no. 8, pp. 9938–9946, Apr. 2015.

[102] C. Hoessbacher *et al.*, "Plasmonic Modulator with >170 GHz Bandwidth Demonstrated at 100 GBd NRZ," *Opt. Express*, vol. 25, no. 3, pp. 1762–1768, Feb. 2017.

[103] G. Raybon *et al.*, "Single Carrier High Symbol Rate Transmitter for Data Rates up to 1.0 Tb/s," in *Optical Fiber Communication Conference (OFC)*, 2016, p. Th3A.2.

[104] J. Jiang, P. Chiang, H. Hung, C. Lin, T. Yoon, and J. Lee, "100Gb/s Ethernet Chipsets in 65nm CMOS Technology," in *IEEE International Solid-State Circuits Conference (ISSCC)*, 2013, pp. 120–122.

[105] D. M. Kuchta, "Directly Modulated VCSELs at $\geq$ 50 Gb/s for Short Reach Data Communications," in *2016 IEEE Optical Interconnects Conference (OI)*, 2016, p. MD1.

[106] A. Mekis *et al.*, "High-Speed Silicon Photonics Optical Transceivers," in *2015 IEEE Summer Topicals Meeting Series (SUM)*, 2015, pp. 23–24.

[107] J. D. Ambrosia and S. G. Kipp, "The 2015 Ethernet Roadmap," *Ethernet Alliance White Paper*. 2015.

[108] Yichuan Yu *et al.*, "80 Gb/s ETDM Transmitter with a Traveling-Wave Electroabsorption Modulator," in *Optical Fiber Communication Conference (OFC)*, 2005, p. OWE1.

[109] S. Lange, R. Kaiser, M. Gruner, M. Hamacher, K.-O. Velthaus, and M. Schell, "Low Switching Voltage InP-Based Travelling Wave Electrode Mach-Zehnder Modulator Monolithically Integrated with DFB-Laser for 60 Gb/s NRZ," in *Optical Fiber Communication Conference (OFC)*, 2015, p. Th4E.1.

[110] Y. Ogiso *et al.*, "100 Gb/s and 2 V Vπ InP Mach-Zehnder Modulator with an N-I-P-N Heterostructure," *Electron. Lett.*, vol. 52, no. 22, pp. 1866–1867, Oct. 2016.

[111] S. Lange *et al.*, "100 GBd Intensity Modulation and Direct Detection with an InP-Based Monolithic DFB Laser Mach-Zehnder Modulator," *Opt. Fiber Commun. Conf.*, p. Th5C.5, 2017.

[112] D. Feng *et al.*, "High Speed GeSi Electro-Absorption Modulator at 1550 Nm Wavelength on SOI Waveguide," *Opt. Express*, vol. 20, no. 20, pp. 22224–22232, Sep. 2012.

[113] A. Srinivasan *et al.*, "50Gb/s C-Band GeSi Waveguide Electro-Absorption Modulator," in *Optical Fiber Communication Conference (OFC)*, 2016, p. Tu3D.7.

[114] J. Verbist *et al.*, "First Real-Time 100-Gb/s NRZ-OOK Transmission over 2 Km with a Silicon Photonic Electro-Absorption Modulator," in *Optical Fiber Communication Conference (OFC)*, 2017, p. Th5C.4.

[115] H. Xu *et al.*, "High-Speed Silicon Modulator with Band Equalization," *Opt. Lett.*, vol. 39, no. 16, pp. 4839–4842, 2014.

[116] X. Xiao, M. Li, Z. Li, L. Wang, Q. Yang, and S. Yu, "Substrate Removed Silicon Mach-Zehnder Modulator for High Baud Rate Optical Intensity Modulations," in *Optical Fiber Communication Conference (OFC)*, 2016, p. Th4H.5.

[117] X. Xiao, M. Li, L. Wang, D. Chen, Q. Yang, and S. Yu, "High Speed Silicon Photonic Modulators," in *Optical Fiber Communication Conference (OFC)*, 2017, p. Tu2H.1.

[118] M. Lauermann *et al.*, "40 GBd 16QAM Signaling at 160 Gb/s in a Silicon-Organic Hybrid Modulator," *J. Light. Technol.*, vol. 33, no. 6, pp. 1210–1216, Mar. 2015.

[119] M. Lauermann *et al.*, "Low-Power Silicon-Organic Hybrid (SOH) Modulators for Advanced Modulation Formats," *Opt. Express*, vol. 22, no. 24, pp. 29927–29936, Nov. 2014.

[120] J. Pfeifle, L. Alloatti, W. Freude, J. Leuthold, and C. Koos, "Silicon-Organic Hybrid Phase Shifter Based on a Slot Waveguide with a Liquid-Crystal Cladding.," *Opt. Express*, vol. 20, no. 14, pp. 15359–76, Jul. 2012.

[121] W. Hartmann *et al.*, "100 Gbit/s OOK Using a Silicon-Organic Hybrid (SOH) Modulator," in *European Conference on Optical Communication (ECOC)*, 2015, p. PDP.1.4.

[122] H. Zwickel *et al.*, "120 Gbit/s PAM-4 Signaling Using a Silicon-Organic Hybrid (SOH) Mach-Zehnder Modulator," in *European Conference on Optical Communication (ECOC)*, 2016, pp. 1058–1060.

[123] S. Wolf *et al.*, "Silicon-Organic Hybrid (SOH) IQ Modulator for 100 GBd 16QAM Operation," in *Optical Fiber Communication Conference (OFC)*, 2017, p. Th5C.1.

[124] H. Sato *et al.*, "Low Driving Voltage Mach-Zehnder Interference Modulator Constructed from an Electro-Optic Polymer on Ultra-Thin Silicon with a Broadband Operation," *Opt. Express*, vol. 25, no. 2, pp. 768–775, Jan. 2017.

[125] N. Ishikura, K. Goi, H. Zhu, M. Illarionov, H. Ishihara, and A. Oka, "Transmission Characteristics of 32-Gbaud PDM IQ Monolithic Silicon Modulator Operating with 2-V PPD Drive Voltage," in *European Conference on Optical Communication (ECOC)*, 2016, p. W.2.E.4.

[126] D. Korn *et al.*, "Lasing in Silicon–organic Hybrid Waveguides," *Nat. Commun.*, vol. 7, p. 10864, Mar. 2016.

[127] B. Snyder, B. Corbett, and P. O'Brien, "Hybrid Integration of the Wavelength-Tunable Laser With a Silicon Photonic Integrated Circuit," *J. Light. Technol.*, vol. 31, no. 24, pp. 3934–3942, Dec. 2013.

[128] N. Lindenmann *et al.*, "Photonic Wire Bonding: A Novel Concept for Chip-Scale Interconnects.," *Opt. Express*, vol. 20, no. 16, pp. 17667–77, Jul. 2012.

[129] N. Lindenmann *et al.*, "Connecting Silicon Photonic Circuits to Multicore Fibers by Photonic Wire Bonding," *J. Light. Technol.*, vol. 33, no. 4, pp. 755–760, Feb. 2015.

[130] M. R. Billah *et al.*, "8-Channel 448 Gbit/s Silicon Photonic Transmitter Enabled by Photonic Wire Bonding," in *Optical Fiber Communication Conference (OFC)*, 2017, p. Th5D.6.

[131] M. R. Billah *et al.*, "Multi-Chip Integration of Lasers and Silicon Photonics by Photonic Wire Bonding," in *Conference on Lasers and Electro-Optics (CLEO)*, 2015, p. STu2F.2.

[132] P.-I. C. Dietrich *et al.*, "Lenses for Low-Loss Chip-to-Fiber and Fiber-to-Fiber Coupling Fabricated by 3D Direct-Write Lithography," in *Conference on Lasers and Electro-Optics*, 2016, p. SM1G.4.

[133] F. Chang, K. Onohara, and T. Mizuochi, “Forward Error Correction for 100 G Transport Networks,” *IEEE Commun. Mag.*, vol. 48, no. 3, pp. S48–S55, Mar. 2010.

[134] H. Kim and A. H. Gnauck, “Chirp Characteristics of Dual-Drive Mach-Zehnder Modulator with a Finite DC Extinction Ratio,” *IEEE Photonics Technol. Lett.*, vol. 14, no. 3, pp. 298–300, 2002.

[135] F. Koyama and K. Oga, “Frequency Chirping in External Modulators,” *J. Light. Technol.*, vol. 6, no. 1, pp. 87–93, 1988.

[136] L. Chen, P. Dong, and Y.-K. Chen, “Chirp and Dispersion Tolerance of a Single-Drive Push-Pull Silicon Modulator at 28 Gb/s,” *IEEE Photonics Technol. Lett.*, vol. 24, no. 11, pp. 936–938, Jun. 2012.

[137] D. Janner, D. Tulli, M. Belmonte, and V. Pruneri, “Waveguide Electro-Optic Modulation in Micro-Engineered LiNbO 3,” *J. Opt. A Pure Appl. Opt.*, vol. 10, no. 10, p. 104003, Oct. 2008.

[138] M. M. Strasser, M. Pfennigbauer, M. Pauer, and P. J. Winzer, “Experimental Verification of Optimum Filter Bandwidths in Direct-Detection (N)RZ Receivers Limited by Optical Noise,” in *LEOS 2001; 14th Annual Meeting of the IEEE Lasers and Electro-Optics Society*, 2001, pp. 485–486.

[139] Mellanox, “OpenOptics Design Guide.” [Online]. Available: www.openopticsmsa.org/pdf/Open_Optics_Design_Guide.pdf.

[140] Z. Wang and A. Ghiasi, “FEC Tradeoffs and Analyses for 100G Optical Networking,” 2012.

[141] J. Bauwelinck *et al.*, “On-Chip Transmitter and Receiver Front-Ends for Ultra- Broadband Wired and Optical-Fiber Communications,” in *Optical Fiber Communication Conference (OFC)*, 2016, p. Tu3B.1.

[142] C. Sun *et al.*, “Single-Chip Microprocessor That Communicates Directly Using Light,” *Nature*, vol. 528, pp. 534–538, Dec. 2015.

[143] D. M. Kuchta *et al.*, “A 71-Gb/s NRZ Modulated 850-Nm VCSEL-Based Optical Link,” *IEEE Photonics Technol. Lett.*, vol. 27, no. 6, pp. 577–580, Mar. 2015.

[144] S. Huang *et al.*, "Highly Efficient Electro-Optic Polymers through Improved Poling Using a Thin TiO2-Modified Transparent Electrode," *Appl. Phys. Lett.*, vol. 96, no. 24, p. 243311, Jun. 2010.

[145] D. Liang and J. E. Bowers, "Recent Progress in Lasers on Silicon," *Nat. Photonics*, vol. 4, no. 8, pp. 511–517, 2010.

[146] G. de Valicourt, A. Verdier, R. Brenot, Y.-K. Chen, and P. Dong, "Integrated Ultra-Wide Band Wavelength-Tunable Hybrid External Cavity Silicon-Based Laser," in *Optical Fiber Communication Conference (OFC)*, 2017, p. Th5C.6.

[147] H. Miura *et al.*, "High Thermal Stability 40 GHz Electro-Optic Polymer Modulators," *Opt. Express*, vol. 25, no. 23, pp. 28643–28649, 2017.

[148] U. Hölzle, "Plenary Talk," in *Optical Fiber communication Conference (OFC)*, 2017.

[149] M. Lauermann *et al.*, "Multi-Channel, Widely-Tunable Coherent Transmitter and Receiver PICs Operating at 88Gbaud/16-QAM," in *Optical Fiber Communication Conference (OFC)*, 2017, p. Th5C.2.

[150] Y. Ogiso *et al.*, "Ultra-High Bandwidth InP IQ Modulator with 1.5 V Vπ," in *European Conference on Optical Communication (ECOC)*, 2016, pp. 355–357.

[151] J. Orcutt *et al.*, "Monolithic Silicon Photonics at 25Gb/s," in *Optical Fiber Communication Conference (OFC)*, 2016, p. Th4H.1.

[152] C. R. Doerr *et al.*, "O, E, S, C, and L Band Silicon Photonics Coherent Modulator/Receiver," in *Optical Fiber Communication Conference (OFC)*, 2016, p. Th5C.4.

[153] C. R. Doerr *et al.*, "Silicon Photonics Coherent Transceiver in a Ball-Grid Array Package," in *Optical Fiber Communication Conference (OFC)*, 2017, p. Th5D.5.

[154] S. S. Azadeh *et al.*, "Low Vπ Silicon Photonics Modulators with Highly Linear Epitaxially Grown Phase Shifters," *Opt. Express*, vol. 23, no. 18, pp. 23526–23550, 2015.

[155] Y. Fang *et al.*, "Silicon IQ Modulator Based 480km 80x453.2Gb/s PDM-eOFDM Transmission on 50GHz Grid with SSMF and EDFA-Only Link," in *Optical Fiber Communication Conference (OFC)*, 2015, p. M3G.5.

[156] M. R. Watts, W. A. W. A. Zortman, S. Member, D. C. D. C. Trotter, R. W. R. W. Young, and A. L. A. L. Lentine, "Low-Voltage, Compact, Depletion-Mode, Silicon Mach-Zehnder Modulator," *IEEE J. Sel. Top. Quantum Electron.*, vol. 16, no. 1, pp. 159–164, 2010.

[157] Chi Xiong, D. Gill, J. Proesel, J. Orcutt, W. Haensch, and W. M. J. Green, "A Monolithic 56 Gb/s CMOS Integrated Nanophotonic PAM-4 Transmitter," in *2015 IEEE Optical Interconnects Conference (OI)*, 2015, pp. 16–17.

[158] P. Dong, C. Xie, L. Buhl, Y. Chen, J. Sinsky, and G. Raybon, "Silicon In-Phase/Quadrature Modulator with On-Chip Optical Equalizer," *J. Light. Technol.*, vol. 33, no. 6, pp. 1191–1196, 2015.

[159] C. Koos, J. Brosi, M. Waldow, W. Freude, and J. Leuthold, "Silicon-on-Insulator Modulators for next-Generation 100 Gbit/s-Ethernet," in *European Conference on Optical Communication (ECOC)*, 2007, p. P056.

[160] C. Bosshard, K. Sutter, R. Schlesser, and P. Günter, "Electro-Optic Effects in Molecular Crystals," *J. Opt. Soc. Am. B*, vol. 10, no. 5, pp. 867–885, May 1993.

[161] S. Wolf *et al.*, "Silicon-Organic Hybrid (SOH) Mach-Zehnder Modulators for 100 Gbit/s on-off Keying," *Sci. Rep.*, vol. 8, p. 2598, 2018.

[162] S. Wolf *et al.*, "An Energy-Efficient 252 Gbit/s Silicon-Based IQ-Modulator," in *Optical Fiber Communication Conference (OFC)*, 2016, p. Th3J.2.

[163] "Institute of Microelectronics, Agency for Science, Technology and Research (A*STAR)." [Online]. Available: https://www.a-star.edu.sg.

[164] R. Ding *et al.*, "Low-Loss Asymmetric Strip-Loaded Slot Waveguides in Silicon-on-Insulator," *Appl. Phys. Lett.*, vol. 98, p. 233303, 2011.

[165] "Keysight Technologies 89600 VSA Software." .

[166] K. Schuh *et al.*, "100 GSa/s BiCMOS DAC Supporting 400 Gb/s Dual Channel Transmission," in *European Conference on Optical Communication (ECOC)*, 2016, pp. 37–39.

[167] D. Chang *et al.*, "FPGA Verification of a Single QC-LDPC Code for 100 Gb/s Optical Systems without Error Floor down to BER of 10^-15," in *Optical Fiber Communication Conference (OFC)*, 2011, p. OTuN2.

[168] K. Cushon, P. Larsson-Edefors, and P. Andrekson, "Low-Power 400-Gbps Soft-Decision LDPC FEC for Optical Transport Networks," *J. Light. Technol.*, vol. 34, no. 18, pp. 4304–4311, 2016.

[169] L. M. Zhang and F. R. Kschischang, "Staircase Codes With 6% to 33% Overhead," *J. Light. Technol.*, vol. 32, no. 10, pp. 1999–2002, 2014.

[170] S. J. Savory, "Digital Filters for Coherent Optical Receivers," *Opt. Express*, vol. 16, no. 2, pp. 804–817, 2008.

[171] A. Konczykowska, J.-Y. Dupuy, and F. Jorge, "42 GBd 3-Bit Power-DAC for Optical Communications with Advanced Modulation Formats in InP DHBT," *Electron. Lett.*, vol. 47, no. 6, pp. 389–390, 2011.

[172] D. Brunina, X. Zhu, K. Padmaraju, L. Chen, M. Lipson, and K. Bergman, "10-Gb/s WDM Optically-Connected Memory System Using Silicon Microring Modulators," in *European Conference on Optical Communication (ECOC)*, 2012, p. Mo.2.A.5.

[173] M. Secondini, E. Forestieri, and F. Cavaliere, "Novel Optical Modulation Scheme for 16-QAM Format with Quadrant Differential Encoding," in *Photonics in Switching*, 2009.

[174] G.-W. Lu *et al.*, "40-Gbaud 16-QAM Transmitter Using Tandem IQ Modulators with Binary Driving Electronic Signals.," *Opt. Express*, vol. 18, no. 22, pp. 23062–9, Oct. 2010.

[175] A. Chiba, T. Sakamoto, T. Kawanishi, and K. Higuma, "16-Level Quadrature Amplitude Modulation by Monolithic Quad-Parallel Mach-Zehnder Optical Modulator," *Electron. Lett.*, vol. 46, no. 3, pp. 220–222, 2010.

[176] A. Shastri *et al.*, "Experimental Demonstration of Ultra-Low-Power Single Polarization 56 Gb/s QAM-16 Generation without DAC Using CMOS Photonics," in *European Conference on Optical Communication (ECOC)*, 2014, p. PD.2.5.

[177] S. Wolf *et al.*, "10 GBd SOH Modulator Directly Driven by an FPGA without Electrical Amplification," in *European Conference on Optical Communication (ECOC)*, 2014, p. Mo.4.5.4.

[178] S. Wolf *et al.*, "DAC-Less and Amplifier-Less Generation and Transmission of 16QAM Signals Using a Sub-Volt Silicon Photonic Modulator," in *European Conference on Optical Communication (ECOC)*, 2014, p. PD.4.5.

[179] J. Leuthold *et al.*, "Silicon-Organic Hybrid Electro-Optical Devices," *IEEE J. Sel. Top. Quantum Electron.*, vol. 19, no. 6, p. 3401413, Nov. 2013.

[180] M. Lauermann *et al.*, "16QAM Silicon-Organic Hybrid (SOH) Modulator Operating with 0.6 Vpp and 19 fJ/bit at 112 Gbit/s," in *CLEO: Science and Innovations*, 2014, p. SM2G.6.

[181] R. Palmer *et al.*, "High-Speed Silicon-Organic Hybrid (SOH) Modulator with 1, 6 fJ/bit and 180 pm/V in-Device Nonlinearity," in *European Conference on Optical Communication (ECOC)*, 2013, p. We.3.B.3.

[182] D. Vermeulen *et al.*, "High-Efficiency Fiber-to-Chip Grating Couplers Realized Using an Advanced CMOS-Compatible Silicon-on-Insulator Platform.," *Opt. Express*, vol. 18, no. 17, pp. 18278–18283, 2010.

[183] P. Dong *et al.*, "Low Loss Shallow-Ridge Silicon Waveguides.," *Opt. Express*, vol. 18, no. 14, pp. 14474–9, Jul. 2010.

[184] M. Vanhoecke *et al.*, "Multi-Level Optical Signal Generation Using a Segmented-Electrode InP IQ-MZM with Integrated CMOS Binary Drivers," in *European Conference on Optical Communication (ECOC)*, 2016, pp. 352–354.

[185] K. Kikuchi, "Fundamentals of Coherent Optical Fiber Communications," *J. Light. Technol.*, vol. 34, no. 1, pp. 157–179, Jan. 2016.

[186] S. J. Savory, "Digital Coherent Optical Receivers: Algorithms and Subsystems," *IEEE J. Sel. Top. Quantum Electron.*, vol. 16, no. 5, pp. 1164–1179, Sep. 2010.

[187] S. J. Savory, G. Gavioli, R. I. Killey, and P. Bayvel, "Electronic Compensation of Chromatic Dispersion Using a Digital Coherent Receiver," *Opt. Express*, vol. 15, no. 5, p. 2120, 2007.

[188] N. Kaneda, J. Lee, and Y.-K. Chen, "Nonlinear Equalizer for 112-Gb/s SSB-PAM4 in 80-Km Dispersion Uncompensated Link," in *Optical Fiber Communication Conference (OFC)*, 2017, p. Tu2D.5.

[189] J. Shi, Y. Zhou, Y. Xu, J. Zhang, J. Yu, and N. Chi, "200-Gbps DFT-S OFDM Using DD-MZM-Based Twin-SSB With a MIMO-Volterra Equalizer," *IEEE Photonics Technol. Lett.*, vol. 29, no. 14, pp. 1183–1186, Jul. 2017.

[190] E. Giacoumidis *et al.*, "Comparison of DSP-Based Nonlinear Equalizers for Intra-Channel Nonlinearity Compensation in Coherent Optical OFDM," *Opt. Lett.*, vol. 41, no. 11, p. 2509, Jun. 2016.

[191] T. Pfau, S. Hoffmann, and R. Noe, "Hardware-Efficient Coherent Digital Receiver Concept With Feedforward Carrier Recovery for M-QAM Constellations," *J. Light. Technol.*, vol. 27, no. 8, pp. 989–999, Apr. 2009.

[192] A. Viterbi, "Nonlinear Estimation of PSK-Modulated Carrier Phase with Application to Burst Digital Transmission," *IEEE Trans. Inf. Theory*, vol. 29, no. 4, pp. 543–551, Jul. 1983.

[193] I. Fatadin, D. Ives, and S. J. Savory, "Laser Linewidth Tolerance for 16-QAM Coherent Optical Systems Using QPSK Partitioning," *IEEE Photonics Technol. Lett.*, vol. 22, no. 9, pp. 631–633, May 2010.

[194] I. Fatadin, D. Ives, and S. J. Savory, "Blind Equalization and Carrier Phase Recovery in a 16-QAM Optical Coherent System," *J. Light. Technol.*, vol. 27, no. 15, pp. 3042–3049, Aug. 2009.

[195] L. Pakala and B. Schmauss, "Extended Kalman Filtering for Simultaneous Phase and Amplitude Noise Mitigation in WDM Systems," in *2015 17th International Conference on Transparent Optical Networks (ICTON)*, 2015, p. Mo.B1.4.

[196] T. Mizuochi, Y. Miyata, K. Kubo, T. Sugihara, K. Onohara, and H. Yoshida, "Progress in Soft-Decision FEC," in *Optical Fiber Communication Conference (OFC)*, 2011, p. NWC2.

[197] G. Tzimpragos, C. Kachris, I. Djordjevic, M. Cvijetic, D. Soudris, and I. Tomkos, "A Survey on FEC Codes for 100G and Beyond Optical Networks," *IEEE Commun. Surv. Tutorials*, vol. 18, no. 1, pp. 209-221, 2016

[198] A. Ghazisaeidi *et al.*, "Transoceanic Transmission Systems Using Adaptive Multirate FECs," *J. Light. Technol.*, vol. 33, no. 7, pp. 1479-1487, Apr. 2015.

[199] G. Grau and W. Freude, *Optische Nachrichtentechnik – Eine Einführung*, 2005.

[200] A. H. Gnauck *et al.*, "Dispersion Penalty Reduction Using an Optical Modulator with Adjustable Chirp," *IEEE Photonics Technol. Lett.*, vol. 3, no. 10, pp. 916–918, 1991.

[201] S. Walklin and J. Conradi, "Effect of Mach-Zehnder Modulator DC Extinction Ratio on Residual Chirp-Induced Dispersion in 10-Gb/s Binary and AM-PSK Duobinary Lightwave Systems," *IEEE Photonics Technol. Lett.*, vol. 9, no. 10, pp. 1400–1402, 1997.

[202] F. Devaux, Y. Sorel, and J. F. Kerdiles, “Simple Measurement of Fiber Dispersion and of Chirp Parameter of Intensity Modulated Light Emitter,” *J. Light. Technol.*, vol. 11, no. 12, pp. 1937–1940, 1993.

[203] S. Lange *et al.*, “100 GBd Intensity Modulation and Direct Detection With an InP-Based Monolithic DFB Laser Mach–Zehnder Modulator,” *J. Light. Technol.*, vol. 36, no. 1, pp. 97–102, Jan. 2018.

H. Glossary

H.1 List of abbreviations

16QAM	16-state quadrature amplitude modulation
4ASK	4-state amplitude shift keying
4PAM	4-state pulse-amplitude modulation
8ASK	8-state amplitude shift keying
64QAM	64-state quadrature amplitude modulation
8PSK	8-state phase shift keying
AD	Analog-to-digital
Ag	Silver
Al	Aluminum
AOC	Active optical cables
ASE	Amplified spontaneous emission
Au	Gold
AWG	Arbitrary waveform generator
b2b	Back-to-back
BER	Bit error ratio
BOX	Buried oxide
BPD	Balanced photodetector
BPF	Bandpass filter
BPSK	Binary phase shift keying

CMOS	Complementary metal-oxide-semiconductor
Cu	Copper
CW	Continuous wave
DA	Digital-to-analog
DAC	Digital-to-analog converter
DC	Direct current
DCA	Digital communications analyzer
DSP	Digital signal processing
DUV	deep ultraviolet
DQPSK	Differential quadrature phase shift keying
EAM	Electro-absorption modulators
ECL	External cavity laser
EDFA	Erbium-doped fiber amplifier
EO	Electro-optic
EOE	Electro-optic-electric
eOFDM	Electrical orthogonal frequency-division multiplexing
E/O	Electrical-to-optical
ER	Extinction ratio
EVM, EVM_m	Error vector magnitude, EVM_m: normalization to maximum vector of constellation
FEC	Forward error correction
FIR	Finite-impulse response
FPGA	Field-programmable gate array

FWHM	Full-width-half-maximum
G	Ground
GC	Grating coupler
GSG	Ground – signal – ground
HD	Hard-decision
I	In-phase component
ICI	Inter-channel interference
IM/DD	Intensity modulation and direct detection
InP	Indium phosphide
IoT	Internet-of-Things
ISI	Inter-symbol interference
IQ	in-phase/quadrature
$LiNbO_3$	Lithium Niobate
LO	Local oscillator
MIM	Metal-insulator-metal
MMF	Multi-mode fiber
MMI	Multimode interference coupler
MSA	Multi-source agreement
MUX	Multiplexer
MZ	Mach-Zehnder
M-QAM	M-ary quadrature amplitude modulation
MZM	Mach-Zehnder modulator
NGC	Net coding gain

NRZ	Non-return-to-zero
OLED	Organic light-emitting diodes
OMA	Optical modulation analyzer
OOK	On-off keying
O/E	Optical-to-electrical
PAM	Pulse-amplitude modulation
PDF	Probability density function
PDM	Polarization-division multiplexing
PIC	Photonic integrated circuit
PMD	Polarization mode dispersion
POF	Plastic optical fibers
POH	Plasmonic-organic hybrid
PPM	Pulse position modulation
PRBS	Pseudo-random binary sequence
PSK	Phase shift keying
Q	Quadrature component
QAM	Quadrature amplitude modulation
QPSK	Quadrature phase shift keying
RC	Raised-cosine
RF	Radio frequency
ROSA	Receiver optical subassemblies
RRC	Root-raised cosine
Rx	Receiver

SE	Spectral efficiency
SFP	Small form-factor pluggables
SerDes	Serializer/deserializer
Si	Silicon
SiGe	Silicon-germanium
SiP	Silicon photonic
SMF	Single-mode fiber
SNR	Signal-to-noise power ratio
SOH	Silicon organic hybrid
SOI	Silicon on insulator
SOP	State of polarization
SPP	Surface plasmon polarition
TIA	Transimpedance amplifier
TPA	Two-photon absorption
Tx	Transmitter
VCSEL	Vertical-cavity surface emitting laser
VNA	Vector network analyzer
WDM	Wavelength-division multiplexing

H.2 List of mathematical symbols

Greek symbols

α	Chirp Factor
α_{L}	Power loss parameter
β	Roll-off factor of (root) raised-cosine signals
Δ	Difference (e.g. Δn)
$\delta(t)$	Dirac function
δ_{ik}	Kronecker delta
$\delta^{(\mathrm{stat})}$	Extinction ratio (static)
$\delta^{(\mathrm{dyn})}$	Extinction ratio (dynamic)
ε_0	Vacuum permittivity
ε_r	Relative dielectric permittivity
φ, $\varphi(t)$	Phase shift
γ	Field amplitude ratio
Γ	Field interaction factor
λ	Wavelength
λ_0	Vacuum wavelength
η	Quantum efficiency
σ	Standard deviation
$\underline{\chi}^{(n)}$	Response function of order n
$\underline{\tilde{\chi}}^{(n)}$	Electric susceptibility of order n
ω	Angular frequency

ω_0	Angular frequency of the optical carrier
ω_m	Angular frequency of modulating signal

Latin symbols

a	Propagation loss, given in dB/mm
a_k	Complex-valued symbol
a_k^r	Real part of complex-valued symbol
a_k^i	Imaginary part of complex-valued symbol
$a_{k(i)}$	i^{th} symbol in a data signal
$\underline{A}$, $\underline{A}(t)$	Complex-valued amplitude
A, $A(t)$	Real-valued amplitude
A_{slot}	Cross-sectional area of the slot waveguide, h_{Si} x w_{slot}
c	Speed of light in vacuum
C	Capacitance
E, $E(t)$	Electric field
e	Elementary charge (1.6×10^{-19} As)
$\vec{\mathcal{E}}$	Transverse electric mode field
f	Frequency
$\vec{\mathcal{H}}$	Transverse magnetic mode field
$\hbar$	Reduced Planck constant (1.055×10^{-34} Js)
h_{rail}	Height of silicon rails of an SOH modulator
h_{slab}	Height of silicon rails of an SOH modulator
i, $i(t)$	Photo current

k	Propagation constant, $k = 2\pi/\lambda$
k_{norm}	Conversion factor between EVM normalizations (max–avg) in Eqs. (2.25), (2.26)
L	Length
L_{N}	Number of signal levels in Eq. (2.26)
m	Number of bits
M	Number of elements in an alphabet of symbols
$\mathbf{M}_{2\text{x}2}$	Transfer matrix of a 2 × 2 MMI coupler
N	Number of symbols evaluated
n	Refractive index
n_{EO}	Refractive index of the EO material
$p, p(t)$	Time-domain pulse shape
p_{RC}	Time domain pulse shape of signal with RC amplitude spectrum
$\tilde{p}_{\text{RC}}, \tilde{p}_{\text{RRC}}$	RC or RRC amplitude spectrum of pulse shape
p_{b}	Probability
$P, P(t)$	Power
$\vec{P}$	Electric polarization
Q	Quality factor
r	Bit rate (line rate)
r_{ijk}	Electro-optic coefficient, the indices $i, \text{j}, k = \{1,2,3\}$ denote the Cartesian components
R	Resistance

R_S	Symbol rate
R_{slab}	Resistance of the silicon slabs
R_t	Terminating Resistance
S	Sensitivity (Responsivity) of a photodiode
t	Time
T_g	Glass transition temperature
T_S	Symbol duration
u, $u(t)$	Electrical voltage
U_{gate}	Gate voltage
U_m	Modulating voltage (or drive voltage)
U_{pp}	Peak-to-peak voltage
U_π	π-voltage
W_{bit}	Energy per bit
w	Probability density function
w_{rail}	Width of the silicon rails
w_{slot}	Width of the silicon slot
$x(t)$	Time signal
Z	Impedance

Danksagung

Diese Dissertation entstand während meiner Promotion am Institut für Photonik und Quantenelektronik (IPQ) am Karlsruher Institut für Technologie (KIT). An dieser Stelle möchte ich allen Personen danken, die mich in den letzten Jahren unterstützt haben.

Zuallererst möchte ich mich bei meinem Doktorvater Prof. Christian Koos für die Betreuung bedanken. Sein Enthusiasmus und seine Ideen waren außerordentlich motivierend und die gemeinsamen Diskussionen wegweisend. Besonderer Dank gilt Prof. Wolfgang Freude – nicht nur für den wertvollen Input und die gründlichen Diskussionen und Korrekturen während der Promotion, sondern auch für die Unterstützung während des Studiums und als Modellberater. Herrn Prof. Thomas Zwick danke ich für seine Unterstützung während der Promotion und für die Betreuung als Korreferent. Bei Prof. Sebastian Randel möchte ich mich für die Unterstützung in den letzten Monaten bedanken. Die neuen Impulse und ein anderer Blickwinkel waren sehr lehrreich.

Für die Unterstützung bei meiner Arbeit möchte ich mich ganz herzlich bei allen Kollegen bedanken. Danke für die Unterstützung bei den gemeinsamen SOH-Messungen an Heiner Zwickel, Clemens Kieninger, Yasar Kutuvantavida, Matthias Lauermann und Wladick Hartmann. Danke für die Unterstützung im Labor und bei der Auswertung, beim Korrigieren von Manuskripten und vielem mehr an Juned Kemal und Pablo Marin. Danke, Philipp Schindler, für die „Starthilfe" in den ersten Monaten als Doktorand. Allen aktuellen und früheren Doktoranden am IPQ danke ich für die gemeinsame, spannende Zeit. Vielen Dank fürs Kickern, Grillen, oder einfach nur für das gemütliche Feierabendbier!

Herzlich bedanke ich mich bei Bernadette Lehmann, Andrea Riemensperger und Tatiana Gassmann für die Unterstützung aus dem Sekretariat. Danke an unser Team von Technikern, David Guder, Martin Winkeler, Oswald Speck und Marco Hummel für die Unterstützung beim Bau unserer Setups, für die großen und kleinen Reparaturen oder Hilfe bei PC-Problemen. Ohne eure Hilfe hätte vieles nicht geklappt – Danke.

Die Arbeit war eingebunden in die EU-Projekte PhoxTroT und BigPipes und wurde unterstützt von der Karlsruhe School of Optics and Photonics (KSOP). Vielen Dank an meinen KSOP-Mentor Martin Lauer für die Ratschläge auf dem Weg zur Promotion. Ich bedanke mich bei allen Projekt-Partnern für die gemeinsame Arbeit. Besonders danken möchte ich Thorsten Wahlbrink (AMO) und Sander Dorrestein (TE) für die sehr gute Zusammenarbeit. Herzlichen Dank an Michael Königsmann und Thomas Dippon (Keysight) und an Rolf Schmid, Lars Altenhain und Joachim Lutz (Micram) für die gemeinsamen Messungen.

Meinen früheren Betreuern und Begleitern während des Studiums, Rene Schmogrow und David Hillerkuss, möchte ich mich für die spannende und lehrreiche Zeit bedanken und dafür, dass ihr mir die optische Kommunikation schmackhaft gemacht habt. Für die Ratschläge und Inspiration während des Studiums möchte ich mich bei Prof. Jürg Leuthold bedanken.

Meinen Studenten Gregor Ronniger, Matthias Meyer, Sebastian Vogel, Stanislav Frolov, Niko Knauer und Hakan Baytekin danke ich für die gemeinsamen Arbeiten und Messungen.

Zuletzt, aber dafür umso mehr, möchte ich mich bei meinen Eltern Iris und Thomas, bei meinem Bruder Torben und bei meinen Großeltern bedanken. Danke für die Unterstützung in all den letzten Jahren. Danke, dass ihr immer ein offenes Ohr hattet.

List of Publications

Journal Publications

[J1] Schmogrow, R.; Winter, M.; Meyer, M.; Hillerkuss, D.; **Wolf, S.**; Baeuerle, B.; Ludwig, A.; Nebendahl, B.; Ben-Ezra, S.; Meyer, J.; Dreschmann, M.; Hubner, M.; Becker, J.; Koos, C.; Freude, W.; Leuthold, J., "Real-time Nyquist pulse generation beyond 100 Gbit/s and its relation to OFDM," *Opt. Express* **20**, 317-337 (2012)

[J2] Schmogrow, R.; Hillerkuss, D.; **Wolf, S.**; Baeuerle, B.; Winter, M.; Kleinow, P.; Nebendahl, B.; Dippon, T.; Schindler, P. C.; Koos, C.; Freude, W.; Leuthold, J., "512QAM Nyquist sinc-pulse transmission at 54 Gbit/s in an optical bandwidth of 3 GHz," *Opt. Express* **20**, 6439-6447 (2012)

[J3] Hillerkuss, D.; Schmogrow, R.; Meyer, M.; **Wolf, S.**; Jordan, M.; Kleinow, P.; Lindenmann, N.; Schindler, P. C.; Melikyan, A.; Yang, X.; Ben-Ezra, S.; Nebendahl, B.; Dreschmann, M.; Meyer, J.; Parmigiani, F.; Petropoulos, P.; Resan, B.; Oehler, A.; Weingarten, K.; Altenhain, L.; Ellermeyer, T.; Möller, M.; Hubner, M.; Becker, J.; Koos, C.; Freude, W.; Leuthold, J., "Single-Laser 32.5 Tbit/s Nyquist WDM Transmission," *J. Opt. Commun. Netw.* **4**, 715-723 (2012)

[J4] Schindler, P. C.; Schmogrow, R.; **Wolf, S.**; Baeuerle, B.; Nebendahl, B.; Koos, C.; Freude, W.; Leuthold, J., "Full flex-grid asynchronous multiplexing demonstrated with Nyquist pulse-shaping," *Opt. Express* **22**, 10923-10937 (2014)

[J5] Weimann, C.; Schindler, P. C.; Palmer, R.; **Wolf, S.**; Bekele, D.; Korn, D.; Pfeifle, J.; Koeber, S.; Schmogrow, R.; Alloatti, L.; Elder, D. L.; Yu, H.; Bogaerts, W.; Dalton, L. R.; Freude, W.; Leuthold, J.; Koos, C., "Silicon-organic hybrid (SOH) frequency comb sources for terabit/s data transmission," *Opt. Express* **22**, 3629-3637 (2014)

[J6] Lauermann, M.; **Wolf, S.**; Schindler, P. C.; Palmer, R.; Koeber, S.; Korn, D.; Alloatti, L.; Wahlbrink, T.; Bolten, J.; Waldow, M.; Koenigsmann, M.; Kohler, M.; Malsam, D.; Elder, D. L.; Johnston, P. V.; Phillips-Sylvain, N.; Sullivan, P. A.; Dalton, L. R.; Leuthold, J.; Freude, W.; Koos, C., "40 GBd 16QAM Signaling at 160 Gbit/s in a Silicon-Organic Hybrid (SOH) Modulator," *J. Lightwave Technol.* **33**, 1210-1216 (2015), (invited)

[J7] **Wolf, S.**; Lauermann, M.; Schindler, P. C.; Ronniger, G.; Geistert, K.; Palmer, R.; Koeber, S.; Bogaerts, W.; Leuthold, J.; Freude, W.; Koos, C., "DAC-less Amplifier-less Generation and Transmission of QAM Signals Using Sub-Volt Silicon-organic Hybrid Modulators," *J. Lightwave Technol.* **33**, 1425-1432 (2015), (invited)

[J8] Melikyan, A.; Koehnle, K.; Lauermann, M.; Palmer, R.; Koeber, S.; Muehlbrandt, S.; Schindler, P. C.; Elder, D. L.; **Wolf, S.**; Heni, W.; Haffner, C.; Fedoryshyn, Y.; Hillerkuss, D.; Sommer, M.; Dalton, L. R.; Thourhout, D. V.; Freude, W.; Kohl, M.; Leuthold, J.; Koos, C., "Plasmonic-organic hybrid (POH) modulators for OOK and BPSK signaling at 40 Gbit/s," *Opt. Express* **23**, 9938-9946 (2015)

[J9] Schindler, P. C.; Agmon, A.; **Wolf, S.**; Bonk, R.; Meder, L.; Meltsin, M.; Ludwig, A.; Schmogrow, R.; Dreschmann, M.; Meyer, J.; Becker, J.; Nazarathy, M.; Ben-Ezra, S.; Pfeiffer, T.; Freude, W.; Leuthold, J.; Koos, C., "Ultra-Dense, Single-Wavelength DFT-Spread OFDMA PON With Laserless 1.2 Gb/s ONU Ready for Silicon Photonics Integration," *J. Lightwave Technol.* **33**, 1650-1659 (2015), (invited)

[J10] Ludwig, A.; Schulz, M.-L.; Schindler, P. C.; **Wolf, S.**; Koos, C.; Freude, W.; Leuthold, J., "Stacked modulation formats enabling highest-sensitivity optical free-space links," *Opt. Express* **23**, 21942-21957 (2015)

[J11] Koos, C.; Leuthold, J.; Freude, W.; Kohl, M.; Dalton, L. R.; Bogaerts, W.; Giesecke, A. L.; Lauermann, M.; Melikyan, A.; Koeber, S.; **Wolf, S.**; Weimann, C.; Muehlbrandt, S.; Koehnle, K.; Pfeifle, J.; Hartmann, W.; Kutuvantavida, Y.; Ummethala, S.; Palmer, R.; Korn, D.; Alloatti, L.; Schindler, P. C.; Elder, D. L.; Wahlbrink, T.; Bolten, J., "Silicon-Organic Hybrid (SOH) and Plasmonic-Organic Hybrid (POH) Integration," *J. Lightwave Technol.* **34**, 1-13 (2016), (invited)

[J12] Lauermann, M.; **Wolf, S.**; Hartmann, W.; Palmer, R.; Kutuvantavida, Y.; Zwickel, H.; Bielik, A.; Altenhain, L.; Lutz, J.; Schmid, R.; Wahlbrink, T.; Bolten, J.; Giesecke, A. L.; Freude, W.; Koos, C., "Generation of 64 GBd 4ASK signals using a silicon-organic hybrid modulator at 80°C," *Opt. Express* **24**, 9389-9396 (2016)

[J13] Muehlbrandt, S.; Melikyan, A.; Harter, T.; Koehnle, K.; Muslija, A.; Vincze, P.; **Wolf, S.**; Jakobs, P. -J.; Fedoryshyn, Y.; Freude, W.; Leuthold, J.; Koos, C., Kohl, M., "Silicon-plasmonic internal-photoemission detector for 40 Gbit/s data reception," *Optica* **3**, 741-747 (2016)

[J14] Kemal, J. N.; Pfeifle, J.; Marin-Palomo, P.; Gutierrez Pascual, M. D.; **Wolf, S.**; Smyth, F.; Freude, W.; Koos, C., "Multi-wavelength coherent transmission using an optical frequency comb as a local oscillator," *Opt. Express* **24**, 25432-25445 (2016)

[J15] Marin-Palomo, P.; Kemal, J. N.; Karpov, M.; Kordts, A.; Pfeifle, J.; Pfeiffer, M.; Trocha, P.; **Wolf, S.**; Brasch, V.; Anderson, M. H.; Rosenberger, R.; Vijayan, K.; Freude, W.; Kippenberg, T. J.; Koos, C., "Microresonator-based solitons for massively parallel coherent optical communications," *Nature* **546**, 274–279 (2017)

[J16] Heni, W.; Kutuvantavida, Y.; Haffner, C.; Zwickel, H.; Kieninger, C.; **Wolf, S.**; Lauermann, M.; Fedoryshyn, Y.; Tillack, A. F.; Johnson, L. E.; Elder, D. L.; Robinson, B. H.; Freude, W.; Koos, C.; Leuthold, J.; Dalton, L. R., "Silicon–Organic and Plasmonic–Organic Hybrid Photonics," *ACS Photonics* **4**, 1576-1590 (2017)

[J17] Zwickel, H.; **Wolf, S.**; Kieninger, C.; Kutuvantavida, Y.; Lauermann, M.; De Keulenaer, T.; Vyncke, A.; Vaernewyck, R.; Luo, J.; Jen, A. K.-Y.; Freude, W.; Bauwelinck, J.; Randel, S.; Koos, C., "Silicon-organic hybrid (SOH) modulators for intensity-modulation / direct-detection links with line rates of up to 120 Gbit/s," *Opt. Express* **25**, 23784-23800 (2017)

[J18] Lange, S.; **Wolf, S.**; Lutz, J.; Altenhain, L.; Schmid, R.; Kaiser, R.; Schell, M.; Koos, C.; Randel, S., "100 GBd Intensity Modulation and Direct Detection with an InP-based Monolithic DFB Laser Mach-Zehnder Modulator," *J. Lightwave Technol.* **36**, 97-102 (2018)

[J19] **Wolf, S.**; Zwickel, H.; Hartmann, W.; Lauermann, M.; Kutuvantavida, Y.; Kieninger, C.; Altenhain, L.; Schmid, R.; Luo, J.; Jen, A. K.-Y.; Randel, S.; Freude, W.; Koos, C., "Silicon-Organic Hybrid (SOH) Mach-Zehnder Modulators for 100 Gbit/s On-Off Keying," *Sci. Rep.* **8**, 2598 (2018)*; arXiv preprint: 1709.01793* (2017)

[J20] **Wolf, S.**; Zwickel, H.; Kieninger, C.; Hartmann, W.; Kutuvantavida, Y.; Lauermann, M.; Freude, W.; Randel, S.; Koos, C., "400 Gbit/s optical data generation with a silicon-organic hybrid (SOH) modulator," *Opt. Express* **26**, 220-232 (2018)*; arXiv preprint: 1710.00056* (2017)

[J21] Trocha, P.; Karpov, M.; Ganin, D.; Pfeiffer, M. H. P.; Kordts, A.; **Wolf, S.**; Krockenberger, J.; Marin-Palomo, P.; Weimann, C.; Randel, S.; Freude, W.; Kippenberg, T. J.; Koos, C., "Ultrafast optical ranging using microresonator soliton frequency combs," *Science* **359**, 887-891 (2018)

Conference Publications

[C1] Schmogrow, R.; Meyer, M.; **Wolf, S.**; Nebendahl, B.; Hillerkuss, D.; Baeuerle, B.; Dreschmann, M.; Meyer, J.; Hubner, M.; Becker, J.; Koos, C.; Freude, W.; Leuthold, J., "150 Gbit/s Real-Time Nyquist Pulse Transmission Over 150 km SSMF Enhanced by DSP with Dynamic Precision," *Optical Fiber Communication Conference (OFC'12), Los Angeles (CA), USA, March 06-10*, paper OM2A.6 *Optical Society of America (OSA)* (2012)

[C2] Schmogrow, R.; **Wolf, S.**; Baeuerle, B.; Hillerkuss, D.; Nebendahl, B.; Koos, C.; Freude, W.; Leuthold, J., "Nyquist Frequency Division Multiplexing for Optical Communications," *Conference on Lasers and Electro-Optics (CLEO 2012), May 08-12*, CTh1H.2 *Optical Society of America* (2012), (invited)

[C3] Hillerkuss, D.; Schmogrow, R.; Meyer, M.; **Wolf, S.**; Jordan, M.; Kleinow, P.; Lindenmann, N.; Schindler, P. C.; Melikyan, A.; Dreschmann, M.; Meyer, J.; Becker, J.; Koos, C.; Freude, W.; Leuthold, J., "Single-laser 32.5 Tbit/s Nyquist-WDM," *International Conference on Photonics in Switching (PS'12), Ajaccio (Corsica), France, 11-14 September* (2012), (invited)

[C4] Weimann, C.; Meier, D.; **Wolf, S.**; Schleitzer, Y.; Totzeck, M.; Heinrich, A.; Hoeller, F.; Leuthold, J.; Freude, W.; Koos, C., "Fast high-precision distance measurements on scattering technical surfaces using frequency combs," *Conf. on Lasers and Electro-Optics (CLEO'13), San Jose (CA), USA, June 9-14*, CTu2I.3 *Optical Society of America* (2013)

[C5] Freude, W.; Schmogrow, R.; Schindler, P. C.; **Wolf, S.**; Nebendahl, B.; Koos, C.; Leuthold, J., "Polarisation demultiplexing in coherent receivers with real-time digital signal processing," *15th International Conference on Transparent Optical Networks (ICTON), 2013, June 23-27*, Paper Tu.B1.1, Cartagena, Spain, June (2013), (invited)

[C6] Weimann, C.; **Wolf, S.**; Korn, D.; Palmer, R.; Koeber, S.; Schmogrow, R.; Schindler, P. C.; Alloatti, L.; Ludwig, A.; Heni, W.; Bekele, D.; Elder, D. L.; Yu, H.; Bogaerts, W.; Dalton, L. R.; Freude, W.; Leuthold, J.; Koos, C., "Silicon-organic hybrid (SOH) frequency comb source for data transmission at 784 Gbit/s," *39th European Conference and Exhibition on Optical Communication (ECOC 2013), September 22-26*, paper Th.2.B.1, London, UK, Sept (2013)

[C7] Weimann, C.; **Wolf, S.**; Meier, D.; Schleitzer, Y.; Totzeck, M.; Heinrich, A.; Hoeller, F.; Freude, W.; Koos, C., "Fast high-precision distance measurements with electro-optic frequency combs," *115th Annual DGaO Conference (DGaO'14), June 10-14, Karlsruhe, Germany*, talk A1 *Deutsche Gesellschaft für angewandte Optik e.V. (DGaO)* (2014)

[C8] Koos, C.; Leuthold, J.; Freude, W.; Dalton, L. R.; Koeber, S.; Palmer, R.; Weimann, C.; Elder, D. L.; Heni, W.; Korn, D.; Pfeifle, J.; **Wolf, S.**; Bekele, D.; Woessner, M.; Alloatti, L.; Schindler, P. C.; Koenig, S., "Femtojoule modulation and frequency comb generation in silicon-organic hybrid (SOH) devices," *Transparent Optical Networks (ICTON), 2014 16th International Conference on*, paper We.C2.1, Graz, Austria, July (2014), (invited)

[C9] Ludwig, A.; Schulz, M.-L.; Schindler, P. C.; Kuder, K.; **Wolf, S.**; Koos, C.; Freude, W.; Leuthold, J., "Stacking Modulation Formats for Highest-Sensitivity," *Photonic Networks and Devices (NETWORKS'14), San Diego (CA), USA, July 13–17*, paper JM2B.2 *Optical Society of America* (2014), (invited)

[C10] **Wolf, S.**; Schindler, P. C.; Ronniger, G.; Lauermann, M.; Palmer, R.; Koeber, S.; Korn, D.; Bogaerts, W.; Leuthold, J.; Freude, W.; Koos, C., "10 GBd SOH modulator directly driven by an FPGA without electrical amplification," *Optical Communication (ECOC), 2014 European Conference on*, paper Mo.4.5.4, Cannes, France, *September 21-25* (2014)

[C11] Lauermann, M.; Schindler, P. C.; **Wolf, S.**; Palmer, R.; Koeber, S.; Korn, D.; Alloatti, L.; Wahlbrink, T.; Bolten, J.; Waldow, M.; Koenigsmann, M.; Kohler, M.; Malsam, D.; Elder, D. L.; Johnston, P. V.; Phillips-Sylvain, N.; Sullivan, P. A.; Dalton, L. R.; Leuthold, J.; Freude, W.; Koos, C., "40 GBd 16QAM modulation at 160 Gbit/s in a silicon-organic hybrid (SOH) modulator," *Optical Communication (ECOC), 2014 European Conference on*, paper We.3.1.3, Cannes, France, *September 21-25* (2014)

[C12] **Wolf, S.**; Lauermann, M.; Ronniger, G.; Schindler, P. C.; Palmer, R.; Koeber, S.; Bogaerts, W.; Leuthold, J.; Freude, W.; Koos, C., "DAC-less and amplifier-less generation and transmission of 16QAM signals using a sub-volt silicon photonic modulator," *European Conference on Optical Communication (ECOC), 2014*, PD 4.5, Cannes, France, *September 21-25* (2014)

[C13] Schindler, P. C.; Agmon, A.; **Wolf, S.**; Bonk, R.; Meder, L.; Meltsin, M.; Ludwig, A.; Becker, J.; Nazarathy, M.; Ben-Ezra, S.; Pfeiffer, T.; Freude, W.; Leuthold, J.; Koos, C., "Ultra-dense, single-wavelength DFT-spread OFDM PON with laserless 1 Gb/s ONU at only 300 MBd per spectral group," *Optical Communication (ECOC), 2014 European Conference on*, paper We.1.6.5, Cannes, France, *September 21-25* (2014)

[C14] Schindler, P. C.; Lauermann, M.; **Wolf, S.**; Korn, D.; Palmer, R.; Koeber, S.; Heni, W.; Ludwig, A.; Schmogrow, R.; Elder, D. L.; Dalton, L. R.; Bogaerts, W.; Yu, H.; Freude, W.; Leuthold, J.; Koos, C., "Ultra-short silicon-organic hybrid (SOH) modulator for bidirectional polarization-independent operation," *Optical Communication (ECOC), 2014 European Conference on*, paper Mo.4.5.5, Cannes, France, *September 21-25* (2014)

[C15] Koos, C.; Freude, W.; Kippenberg, T. J.; Leuthold, J.; Dalton, L. R.; Pfeifle, J.; Weimann, C.; Lauermann, M.; Palmer, R.; Koeber, S.; **Wolf, S.**; Schindler, P. C.; Brasch, V.; Elder, D. L., "Terabit/s optical transmission using chip-scale frequency comb sources," *Optical Communication (ECOC), 2014 European Conference on*, paper Mo.3.4.5, Cannes, France, *September 21-25* (2014), (invited)

[C16] Pfeifle, J.; Shkarban, I.; **Wolf, S.**; Kemal, J. N.; Weimann, C.; Hartmann, W.; Chimot, N.; Joshi, S.; Merghem, K.; Martinez, A.; Weber, M.; Ramdane, A.; Lelarge, F.; Freude, W.; Koos, C., "Coherent terabit communications using a quantum-dash mode-locked laser and self-homodyne detection," *Optical Fiber Communication Conference (OFC'15), Los Angeles (CA), USA, March 22-26*, paper W2A.19 (2015)

[C17] Pfeifle, J.; Watts, R. T.; Shkarban, I.; **Wolf, S.**; Vujicic, V.; Landais, P.; Chimot, N.; Joshi, S.; Merghem, K.; Calo, C.; Weber, M.; Ramdane, A.; Lelarge, F.; Barry, L. P.; Freude, W.; Koos, C., "Simultaneous phase noise reduction of 30 comb lines from a quantum-dash mode-locked laser diode enabling coherent Tbit/s data transmission," *Optical Fiber Communication Conference (OFC'15), Los Angeles (CA), USA, March 22-26*, paper Tu3I.5 (2015)

[C18] Lauermann, M.; **Wolf, S.**; Palmer, R.; Bielik, A.; Altenhain, L.; Lutz, J.; Schmid, R.; Wahlbrink, T.; Bolten, J.; Giesecke, A. L.; Freude, W.; Koos, C., "64 GBd operation of a silicon-organic hybrid modulator at elevated temperature," *Optical Fiber Communication Conference (OFC'15), Los Angeles (CA), USA, March 22–26*, paper Tu2A.5 (2015)

[C19] Koos, C.; Leuthold, J.; Freude, W.; Kohl, M.; Dalton, L. R.; Bogaerts, W.; Giesecke, A. L.; Lauermann, M.; Melikyan, A.; Koeber, S.; **Wolf, S.**; Weimann, C.; Muehlbrandt, S.; Koehnle, K.; Pfeifle, J.; Palmer, R.; Alloatti, L.; Elder, D. L.; Wahlbrink, T.; Bolten, J., "Silicon-Organic Hybrid (SOH) and Plasmonic-Organic Hybrid (POH) Integration," *Optical Fiber Communication Conference (OFC'15), Los Angeles (CA), USA, March 22–26*, paper Tu2A.1 (2015), (invited)

[C20] Lauermann, M.; **Wolf, S.**; Palmer, R.; Koeber, S.; Schindler, P. C.; Wahlbrink, T.; Bolten, J.; Giesecke, A. L.; Koenigsmann, M.; Kohler, M.; Malsam, D.; Elder, D. L.; Dalton, L. R.; Leuthold, J.; Freude, W.; Koos, C., "High-speed and low-power silicon-organic hybrid modulators for advanced modulation formats," *SPIE 9516 Integrated Optics: Physics and Simulations II, Prague, Czech Republic, April 13-16*, 951607-951607-5 (2015), (invited)

[C21] Pfeifle, J.; Kordts, A.; Marin-Palomo, P.; Karpov, M.; Pfeiffer, M.; Brasch, V.; Rosenberger, R.; Kemal, J. N.; **Wolf, S.**; Freude, W.; Kippenberg, T. J.; Koos, C., "Full C and L-Band Transmission at 20 Tbit/s Using Cavity-Soliton Kerr Frequency Combs," *Conf. on Lasers and Electro-Optics (CLEO'15), Postdeadline Paper Digest, San Jose (CA), USA, May 10-15*, paper JTh5C.8. *Optical Society of America (OSA)* (2015)

[C22] Muehlbrandt, S.; Melikyan, A.; Koehnle, K.; Harter, T.; Muslija, A.; Vincze, P.; **Wolf, S.**; Jakobs, P. -J.; Fedoryshyn, Y.; Freude, W.; Leuthold, J.; Koos, C.; Kohl, M., "Plasmonic Internal Photoemission Detectors with Responsivities above 0.12 A/W," *Conf. on Lasers and Electro-Optics (CLEO'15), San Jose (CA), USA, May 10-15*, paper FTh3E.3 *Optical Society of America (OSA)* (2015)

[C23] Melikyan, A.; Koehnle, K.; Lauermann, M.; Palmer, R.; Koeber, S.; Muehlbrandt, S.; Schindler, P. C.; Elder, D. L.; **Wolf, S.**; Heni, W.; Haffner, C.; Fedoryshyn, Y.; Hillerkuss, D.; Sommer, M.; Dalton, L. R.; Van Thourhout, D.; Freude, W.; Kohl, M.; Leuthold, J.; Koos, C., "Plasmonic-organic hybrid (POH) modulators for OOK and BPSK signaling at 40 Gbit/s," *Conf. on Lasers and Electro-Optics (CLEO'15), San Jose (CA), USA, May 10-15*, paper SM1I.1. *Optical Society of America (OSA)* (2015)

[C24] Koos, C.; Freude, W.; Leuthold, J.; Kohl, M.; Dalton, L. R.; Bogaerts, W.; Lauermann, M.; Melikyan, A.; **Wolf, S.**; Weimann, C.; Muehlbrandt, S.; Pfeifle, J.; Koeber, S.; Palmer, R.; Schindler, P. C.; Elder, D. L., "Silicon-organic (SOH) and plasmonic-organic (POH) hybrid integration: Extending the capabilities of silicon photonics and plasmonics," *20th OptoElectronics and Communications Conference (OECC'15), Shanghai, China, June 28 - July 2*, paper JTuE.34 (2015) (invited)

[C25] Koos, C.; Freude, W.; Leuthold, J.; Kohl, M.; Dalton, L. R.; Bogaerts, W.; Lauermann, M.; **Wolf, S.**; Palmer, R.; Koeber, S.; Melikyan, A.; Weimann, C.; Ronniger, G.; Geistert, K.; Schindler, P. C.; Elder, D. L.; Wahlbrink, T.; Bolten, J.; Giesecke, A. L.; Koenigsmann, M.; Kohler, M.; Malsam, D., "Silicon-organic hybrid (SOH) integration for low-power and high-speed signal generation," *17th Intern. Conf. on Transparent Optical Networks (ICTON'15), Budapest, Hungary, July 5-9*, paper We.C5.1 (2015) (invited)

[C26] Freude, W.; Pfeifle, J.; Watts, R. T.; Shkarban, I.; **Wolf, S.**; Vujicic, V.; Landais, P.; Chimot, N.; Joshi, S.; Merghem, K.; Calo, C.; Weber, M.; Ramdane, A.; Lelarge, F.; Barry, L. P.; Koos, C., "Phase-noise compensated carriers from an optical frequency comb allowing terabit transmission," *17th Intern. Conf. on Transparent Optical Networks (ICTON'15), Budapest, Hungary, July 5-9*, paper Tu.B5.1 (2015) (invited)

[C27] Koos, C.; Kippenberg, T. J.; Barry, L. P.; Dalton, L. R.; Ramdane, A.; Lelarge, F.; Leuthold, J.; Freude, W.; Pfeifle, J.; Weimann, C.; Kemal, J. N.; Lauermann, M.; **Wolf, S.**; Shkarban, I.; Koeber, S.; Herr, T.; Brasch, V.; Watts, R. T.; Elder, D. L., "Coherent Terabit Communications using Chip-scale Frequency Comb Sources," *OSA Topical Meeting on Nonlinear Optics (NLO'15), Kauai, Hawaii, USA, July 26-31*, paper NTh2A.1 *Optical Society of America (OSA)* (2015) (invited)

[C28] Kemal, J. N.; Pfeifle, J.; Marin-Palomo, P.; Gutierrez Pascual, M. D.; **Wolf, S.**; Smyth, F.; Freude, W.; Koos, C., "Parallel Multi-Wavelength Intradyne Reception Using an Optical Frequency Comb as a Local Oscillator," *41th European Conf. Opt. Commun. (ECOC'15), Valencia, Spain, September 27 - Oct. 1*, paper P.4.18 (2015)

[C29] Hartmann, W.; Lauermann, M.; **Wolf, S.**; Zwickel, H.; Kutuvantavida, Y.; Luo, J.; Jen, A. K. -Y.; Freude, W.; Koos, C., "100 Gbit/s OOK Using a Silicon - Organic Hybrid (SOH) Modulator," *41th European Conf. Opt. Commun. (ECOC'15), Valencia, Spain, September 27 - Oct. 1*, paper PDP.1.4 (2015) (postdeadline)

[C30] Koos, C.; Freude, W.; Leuthold, J.; Kohl, M.; Dalton, L. R.; Bogaerts, W.; Lauermann, M.; **Wolf, S.**; Weimann, C.; Melikyan, A.; Lindenmann, N.; Billah, M. R.; Muehlbrandt, S.; Koeber, S.; Palmer, R.; Koehnle, K.; Alloatti, L.; Elder, D. L.; Giesecke, A. L.; Wahlbrink, T., "Silicon-organic hybrid (SOH) integration and photonic multi-chip systems: Extending the capabilities of the silicon photonic platform," *28th Annual Conf. of the IEEE Photonics Society (IPC'15), Reston (VA), USA, October 4-8*, paper MF2.1 (2015) (invited)

[C31] **Wolf, S.**; Lauermann, M.; Hartmann, W.; Zwickel, H.; Kutuvantavida, Y.; Koenigsmann, M.; Gruen, M.; Luo, J.; Jen, A. K.-Y.; Freude, W.; Koos, C., "An Energy-Efficient 252 Gbit/s Silicon-Based IQ-Modulator," *Optical Fiber Communication Conference (OFC'16), Anaheim (CA), USA, March 20-24*, paper Th3J.2 (2016)

[C32] Koos, C.; Freude, W.; Leuthold, J.; Dalton, L. R.; **Wolf, S.**; Zwickel, H.; Hoose, T.; Billah, M. R.; Lauermann, M.; Weimann, C.; Hartmann, W.; Melikyan, A.; Lindenmann, N.; Koeber, S.; Palmer, R.; Alloatti, L.; Giesecke, A. L.; Wahlbrink, T., "Silicon-organic hybrid (SOH) integration and photonic multi-chip systems: Technologies for high-speed optical interconnects," *Optical Interconnects Conference (OI'16), Hyatt Regency Mission Bay Spa & Marina, San Diego, California, USA, May 9-11* (2016) (invited)

[C33] Marin-Palomo, P.; Pfeifle, J.; Kemal, J. N.; **Wolf, S.**; Vijayan, K.; Chimot, N.; Martinez, A.; Ramdane, A.; Lelarge, F.; Freude, W.; Koos, C., "8.32 Tbit/s Coherent Transmission Using a Quantum-Dash Mode-Locked Laser Diode," *Conf. on Lasers and Electro-Optics (CLEO'16), San Jose (CA), USA, June 5-10*, paper STh1F.1 (2016)

[C34] Marin-Palomo, P.; Pfeifle, J.; Karpov, M.; Trocha, P.; Rosenberger, R.; Vijayan, K.; **Wolf, S.**; Kemal, J. N.; Kordts, A.; Pfeiffer, M.; Brasch, V.; Freude, W.; Kippenberg, T. J.; Koos, C., "50 Tbit/s Massively Parallel WDM Transmission in C and L Band Using Interleaved Cavity-Soliton Kerr Combs," *Conf. on Lasers and Electro-Optics (CLEO'16), San Jose (CA), USA, June 5-10*, paper STu1G.1 (2016) (invited)

[C35] Melikyan, A.; Koehnle, K.; Lauermann, M.; Palmer, R.; Koeber, S.; Muehlbrandt, S.; Schindler, P. C.; Elder, D. L.; **Wolf, S.**; Sommer, M.; Dalton, L. R.; Thourhout, D. V.; Freude, W.; Kohl, M.; Leuthold, J.; Koos, C., "Plasmonic-organic hybrid (POH) modulators," *21st OptoElectronics and Communications Conference (OECC) held jointly with International Conference on Photonics in Switching (PS'2016), Niigata, Japan, July 3-7*, paper ThE3-1 (2016) (invited)

[C36] Koos, C.; Kippenberg, T. J.; Barry, L. P.; Dalton, L. R.; Ramdane, A.; Lelarge, F.; Freude, W.; Kemal, J. N.; Marin-Palomo, P.; **Wolf, S.**; Trocha, P.; Pfeifle, J.; Weimann, C.; Lauermann, M.; Herr, T.; Brasch, V.; Watts, R. T.; Elder, D. L.; Martinez, A.; Panapakkam, V.; Chimot, N., "Multi-terabit/s transmission using chip-scale frequency comb sources," *18th Intern. Conf. on Transparent Optical Networks (ICTON'16), Trento, Italy, July 10-14*, paper We.A5.1 *IEEE* (2016) (invited)

[C37] Koos, C.; Freude, W.; Dalton, L. R.; Kippenberg, T. J.; Barry, L. P.; Ramdane, A.; Lelarge, F.; **Wolf, S.**; Zwickel, H.; Lauermann, M.; Weimann, C.; Hartmann, W.; Kemal, J. N.; Marin-Palomo, P.; Trocha, P.; Pfeifle, J.; Herr, T.; Brasch, V.; Watts, R. T.; Elder, D. L.; Martinez, A.; Panapakkam, V.; Chimot, N., "Silicon-organic hybrid (SOH) devices and their use in comb-based communication systems," *13th International Conference on Group IV Photonics (GFP'16), Shanghai, China, August 24-26* (2016) (invited)

[C38] Marin-Palomo, P.; Kemal, J. N.; Trocha, P.; **Wolf, S.**; Kordts, A.; Karpov, M.; Pfeiffer, M.; Brasch, V.; Freude, W.; Kippenberg, T. J.; Koos, C., "34.6 Tbit/s WDM transmission using soliton Kerr frequency combs as optical source and local oscillator," *42th European Conf. Opt. Commun. (ECOC'16), Düsseldorf, Germany, September 18-22, p. 415-417* (2016)

[C39] Zwickel, H.; **Wolf, S.**; Kutuvantavida, Y.; Kieninger, C.; Lauermann, M.; Freude, W.; Koos, C., "120 Gbit/s PAM-4 signaling using a silicon-organic hybrid (SOH) Mach-Zehnder modulator," *42th European Conf. Opt. Commun. (ECOC'16), Düsseldorf, Germany, September 18-22, p. 1058-1060* (2016)

[C40] **Wolf, S.**; Hartmann, W.; Lauermann, M.; Zwickel, H.; Kutuvantavida, Y.; Kieninger, C.; Freude, W.; Koos, C., "High-speed silicon-organic hybrid (SOH) modulators," *42th European Conf. Opt. Commun. (ECOC'16), Düsseldorf, Germany, September 18-22, p. 358-360* (2016) (invited)

[C41] Koos, C.; Kippenberg, T. J.; Barry, L. P.; Ramdane, A.; Lelarge, F.; Freude, W.; Marin-Palomo, P.; Kemal, J. N.; Weimann, C.; **Wolf, S.**; Trocha, P.; Pfeifle, J.; Karpov, M.; Kordts, A.; Brasch, V.; Watts, R. T.; Martinez, A.; Panapakkam, V.; Chimot, N., "Chip-scale frequency comb generators for high-speed communications and optical metrology," *SPIE Nonlinear Frequency Generation and Conversion: Materials and Devices XVI, (LASE-SPIE'17), San Francisco (CA), USA, January 28 - February 2*, paper 10090-15 (2017) (invited)

[C42] Koos, C.; Freude, W.; Leuthold, J.; Dalton, L. R.; **Wolf, S.**; Muehlbrandt, S.; Melikyan, A.; Zwickel, H.; Kutuvantavida, Y.; Harter, T.; Lauermann, M.; Elder, D. L., "Nanophotonic modulators and photodetectors using silicon photonic and plasmonic device concepts," *SPIE Physics and Simulation of Optoelectronic Devices XXV (OPTO-SPIE'17), San Francisco (CA), USA, January 28 - February 2*, paper 10098-6 (2017) (invited)

[C43] Zwickel, H.; De Keulenaer, T.; **Wolf, S.**; Kieninger, C.; Kutuvantavida, Y.; Lauermann, M.; Verplaetse, M.; Pierco, R.; Vaernewyck, R.; Vyncke, A.; Yin, X.; Torfs, G.; Freude, W.; Mentovich, E.; Bauwelinck, J.; Koos, C., "100 Gbit/s Serial Transmission Using a Silicon-Organic Hybrid (SOH) Modulator and a Duobinary Driver IC," *Optical Fiber Communication Conference (OFC'17), Los Angeles (CA), USA, March 19-23*, paper W4I.5. *Optical Society of America (OSA)* (2017)

[C44] Kemal, J. N.; Marin-Palomo, P.; Panapakkam, V.; Trocha, P.; **Wolf, S.**; Merghem, K.; Lelarge, F.; Ramdane, A.; Randel, S.; Freude, W.; Koos, C., "WDM Transmission Using Quantum-Dash Mode-Locked Laser Diodes as Multi-Wavelength Source and Local Oscillator," *Optical Fiber Communication Conference (OFC'17), Los Angeles (CA), USA, March 19-23*, paper Th3F.6. (2017)

[C45] **Wolf, S.**; Zwickel, H.; Kieninger, C.; Kutuvantavida, Y.; Lauermann, M.; Lutz, J.; Altenhain, L.; Schmid, R.; Freude, W.; Koos, C.; Randel, S., "Silicon-Organic Hybrid (SOH) IQ Modulator for 100 GBd 16QAM Operation," *Optical Fiber Communication Conference (OFC'17), Los Angeles (CA), USA, March 19-23*, paper Th5C.1 (2017) (postdeadline)

[C46] Lange, S.; **Wolf, S.**; Lutz, J.; Altenhain, L.; Schmid, R.; Kaiser, R.; Koos, C.; Randel, S.; Schell, M., "100 GBd Intensity Modulation and Direct Detection with an InP-based Monolithic DFB Laser Mach-Zehnder Modulator," *Optical Fiber Communication Conference (OFC'17), Los Angeles (CA), USA, March 19-23*, paper Th5C.5 (2017) (postdeadline)

[C47] Harter, T.; Weber, M.; Muehlbrandt, S.; **Wolf, S.**; Kemal, J. N.; Boes, F.; Nellen, S.; Goebel, T.; Giesekus, J.; Zwick, T.; Randel, S.; Freude, W.; Koos, C., "Wireless THz Communications Using Optoelectronic Techniques for Signal Generation and Coherent Reception," *Conf. on Lasers and Electro-Optics (CLEO'17), San Jose (CA), USA, May 14-19*, paper SM3J.2 (2017)

[C48] Kieninger, C.; Kutuvantavida, Y.; Zwickel, H.; **Wolf, S.**; Lauermann, M.; Elder, D. L.; Dalton, L. R.; Freude, W.; Randel, S.; Koos, C., "Record-High In-Device Electro-Optic Coefficient of 359 pm/V in a Silicon-Organic Hybrid (SOH) Modulator," *Conf. on Lasers and Electro-Optics (CLEO'17), San Jose (CA), USA, May 14-19*, paper STu3N.2 (2017)

[C49] Ganin, D.; Trocha, P.; Pfeiffer, M.; Karpov, M.; Kordts, A.; Krockenberger, J.; Marin-Palomo, P.; **Wolf, S.**; Randel, S.; Freude, W.; Kippenberg, T. J.; Koos, C., "Ultrafast Dual-Comb Distance Metrology Using Dissipative Kerr Solitons," *Conf. on Lasers and Electro-Optics (CLEO'17), San Jose (CA), USA, May 14-19*, paper STh4L.6 (2017)

[C50] Billah, M. R.; Kemal, J. N.; Marin-Palomo, P.; Blaicher, M.; Kutuvantavida, Y.; Kieninger, C.; Zwickel, H.; Dietrich, P.-I.; **Wolf, S.**; Hoose, T.; Xu, Y.; Troppenz, U.; Moehrle, M.; Randel, S.; Freude, W.; Koos, C., "Four-Channel 784 Gbit/s Transmitter Module Enabled by Photonic Wire Bonding and Silicon-Organic Hybrid Modulators," *43rd European Conf. Opt. Commun. (ECOC'17), Gothenburg, Sweden, September 17-21*, paper Th.PDP.C.1 (2017) (postdeadline)

[C51] Harter, T.; Adib, M.; **Wolf, S.**; Muehlbrandt, S.; Weber, M.; Blaicher, M.; Boes, F.; Massler, H.; Tessmann, A.; Nellen, S.; Goebel, T.; Giesekus, J.; Walther, M.; Zwick, T.; Freude, W.; Randel, S.; Koos, C., "Wireless Multi-Subcarrier THz Communications Using Mixing in a Photoconductor for Coherent Reception," *IEEE Photonics Conference (IPC2017), Orlando, Florida, USA, October 1-5*, paper MA4.2 (2017)

[C52] Füllner, C.; **Wolf, S.**; Kemal, J.; Lutz, J.; Altenhain, L.; Schmid, R.; Freude, W.; Koos, C.; Randel, S., "Transmission of 80-GBd 16-QAM over 300 km and Kramers-Kronig Reception using a Low-Complexity FIR Hilbert Filter Approximation, " *Optical Fiber Communication Conference (OFC'18), San Diego (CA), USA, March 11-15*, paper W4E.3 (2018)

Curriculum Vitae

Stefan Wolf

born on January 16th 1987
in Aalen, Germany

Nationality: German

Education

02/2013–11/2017	Ph.D. Candidate and Research Associate at the Institute of Photonics and Quantum Electronics at the Karlsruhe Institute of Technology. Field of study: Optical Communications
10/2006–11/2012	Studies at the Karlsruhe Institute of Technology, Germany Major: Electrical Engineering and Information Technology Specialization: Optical Communication Technology Thesis: "Heterodyne Absolute Distance Measurement Using Frequency Combs Generated by Electro-Optic Modulation" Degree: Dipl.-Ing.

School Education

09/1997–07/2006	Kopernikus Gymnasium Wasseralfingen, Aalen, Germany "Abitur" (general university entrance qualification)

Honors and Awards

Corning Outstanding Student Paper Competition:
Honorable Mention at Optical Fiber Communication Conference (OFC) 2016